THE FRONTIERS COLLECTION

THE FRONTIERS COLLECTION

Series Editors:
A.C. Elitzur M. Schlosshauer M.P. Silverman J. Tuszynski R. Vaas H.D. Zeh

The books in this collection are devoted to challenging and open problems at the forefront of modern science, including related philosophical debates. In contrast to typical research monographs, however, they strive to present their topics in a manner accessible also to scientifically literate non-specialists wishing to gain insight into the deeper implications and fascinating questions involved. Taken as a whole, the series reflects the need for a fundamental and interdisciplinary approach to modern science. Furthermore, it is intended to encourage active scientists in all areas to ponder over important and perhaps controversial issues beyond their own speciality. Extending from quantum physics and relativity to entropy, consciousness and complex systems – the Frontiers Collection will inspire readers to push back the frontiers of their own knowledge.

Other Recent Titles

Weak Links
Stabilizers of Complex Systems from Proteins to Social Networks
By P. Csermely

Mind, Matter and the Implicate Order
By P.T.I. Pylkkänen

Particle Metaphysics
A Critical Account of Subatomic Reality
By B. Falkenburg

The Physical Basis of the Direction of Time
By H.D. Zeh

Mindful Universe
Quantum Mechanics and the Participating Observer
By H. Stapp

Decoherence and the Quantum-To-Classical Transition
By M. Schlosshauer

The Nonlinear Universe
Chaos, Emergence, Life
By A. Scott

Symmetry Rules
How Science and Nature Are Founded on Symmetry
By J. Rosen

Quantum Superposition
Counterintuitive Consequences of Coherence, Entanglement, and Interference
By M.P. Silverman

Henry P. Stapp

MIND, MATTER AND QUANTUM MECHANICS

Third Edition

 Springer

Henry P. Stapp
University of California
Lawrence Berkeley National Lab.
1 Cyclotron Rd.
Berkeley CA 94720
USA
e-mail: hpstapp@lbl.gov

Series Editors:

Avshalom C. Elitzur
Bar-Ilan University, Unit of Interdisciplinary Studies, 52900 Ramat-Gan, Israel
email: avshalom.elitzur@weizmann.ac.il

Maximilian A. Schlosshauer
University of Melbourne, Department of Physics, Melbourne, Victoria 3010, Australia
email: m.schlosshauer@unimelb.edu.au

Mark P. Silverman
Trinity College, Dept. Physics, Hartford CT 06106, USA
email: mark.silverman@trincoll.edu

Jack A. Tuszynski
University of Alberta, Dept. Physics, Edmonton AB T6G 1Z2, Canada
email: jtus@phys.ualberta.ca

Rüdiger Vaas
Posener Str. 85, 74321 Bietigheim-Bissingen, Germany
email: Ruediger.Vaas@t-online.de

H. Dieter Zeh
Gaiberger Straße 38, 69151 Waldhilsbach, Germany
email: zeh@uni-heidelberg.de

Cover Figure: Detail from 'The Optiverse', a video of the minimax sphere eversion by John M. Sullivan, George Francis, and Stuart Levy, with original score by Camille Goudeseune. More at http://new.math-uiuc.edu/optiverse

ISBN 978-3-642-43498-3   ISBN 978-3-540-89654-8 (eBook)

DOI 10.1007/978-3-540-89654-8

Frontiers Collection ISSN 1612-3018

Cover design: KuenkelLopka GmbH, Heidelberg

Printed on acid-free paper

9 8 7 6 5 4 3 2 1

springer.com

For Olivia

For Olivia

Preface to the Third Edition

The basic problem in the interpretation of quantum mechanics is to reconcile the quantum features of the mathematics with the fact that our perceptual experiences are described in the language of classical physics. Observed physical objects appear to us to occupy definite locations, and we use the concepts of everyday life, refined by the ideas of nineteenth-century physics, to describe both our procedures for obtaining information about the systems we are studying, and also the data that we then receive, such as the reading of the position of a pointer on a dial. Yet our instruments, and our physical bodies and brains, are in some sense conglomerates of atoms. The individual atoms appear to obey the laws of quantum mechanics, and these laws include rules for combining systems of atomic constituents into larger systems. Insofar as experiments have been able to determine, and these experiments examine systems containing tens of billions of electrons, there is no apparent breakdown of the quantum rules. Yet if we assume that these laws hold all the way up to visible objects such as pointers, then difficulties arise. The state of the pointer would, according to the theory, often have parts associated with the pointer's being located in visibly different places. If we continue to apply the laws right up to, and into, our brains, then our brains, as represented in quantum mechanics, would have parts corresponding to our seeing the pointer in several visibly different locations. Inclusion of the effects of the environment does not remove any of these parts, although it does make it effectively impossible to empirically confirm the simultaneous presence of these different parts.

The orthodox solution to this problem is simply to postulate, as a basic precept of the theory, that our observations are classically describable. This postulate is incorporated into the theory by asserting that any conscious observation will be accompanied by a "collapse of the wave function" or "reduction of the wave packet" that will simply exclude from the prior physically described state all parts that are incompatible with the conscious experience. This prescription works beautifully. When combined with the rule that the probability that this perception will occur is the ratio of the quantum mechanical weighting of the reduced state to the quantum mechanical weighting of the prior state, one gets predictions never known to fail. This ad hoc injection, in association with "consciousness", of "classical" concepts into a theory that is mathematically incompatible with those concepts, is the origin of the mysteriousness of quantum mechanics.

There is mounting evidence from neuroscience that our conscious thoughts are associated with synchronous oscillations in well-separated sites in the brain. This opens the door to a natural way of understanding, simultaneously, both the mind–brain and quantum–classical linkages. Oscillatory motions play a fundamental role in quantum mechanics, and they embody an extremely tight quantum–classical connection. This connection allows the quantum–classical and mind–brain connections to be understood together in a relatively simple and direct way.

Chapters 13 and 14 are new in this edition. Both describe simple models that achieve a simultaneous solution of these two problems. The first paper, entitled "Physicalism Versus Quantum Mechanics", is concerned more with the philosophical aspects, whereas the second, entitled "A Model of the Quantum–Classical and Mind–Brain Connections, and the Role of the Quantum Zeno Effect in the Physical Implementation of Conscious Intent" focuses more on technical matters pertaining to the question of the time scales associated with the quantum-mandated influence of our conscious intentional actions upon our physically described brains. These two papers, and the second one in particular, involve more equations than any of the other papers in the book. But these equations describe properties of simple geometric structures, and the meanings of the equations are described also in geometric terms.

To make room for the new articles without appreciably lengthening the book, the old chapter 5 has been removed. Its content significantly overlapped that of other chapters, so its removal mainly eliminates redundancies.

The two new chapters describe in terms meant to be generally understandable to nonphysicists who are not uncomfortable with mathematics the technical foundations of the approach to the mind–brain connection pursued in this book and further developed in its sequel, the Springer volume *Mindful Universe: Quantum Mechanics and the Participating Observer*.

Berkeley, October 2008   *Henry P. Stapp*

Preface to the Second Edition

I have been besieged by requests for copies of this book, particularly since the publication of *The Mind and the Brain* by Jeffrey Schwartz and Sharon Begley. That book gave a popular-style account of the impact of these quantum-based considerations in psychiatry and neuroscience. This is just one example of the substantial progress that has been made during the decade since the publication of the first edition of *Mind, Matter, and Quantum Mechanics* in understanding the relationship between conscious experience and physical processes in the brain.

Von Neumann's Process I has been identified as the key physical process that accounts, within the framework of contemporary physical theory, for the causal efficacy of directed attention and willful effort. It is now understood how quantum uncertainties in the micro-causal bottom–up physical brain process not only open the door to a consciously controlled top–down process, but also require the presence of this process, at least within the context of pragmatic science.

These new developments fit securely onto the general framework presented in the first edition. They are described in a chapter written for this new edition and entitled "Neuroscience, Atomic Physics, and the Human Person". This chapter integrates the contents of three lectures and a text that I have prepared and delivered during the past year. Those presentations were aimed at four very different audiences, and I have tried to adopt here a style that will make the material accessible to all of those audiences, and hence to a broad readership.

The material covered in that chapter is essentially scientific. The broader ramifications are covered in a second new chapter entitled "Societal Ramifications of the New Scientific Conception of Human Beings".

Berkeley, July 2003   *Henry P. Stapp*

Preface to the First Edition

Nature appears to be composed of two completely different kinds of things: rocklike things and idealike things. The first is epitomized by an enduring rock, the second by a fleeting thought. A rock can be experienced by many of us together, while a thought seems to belong to one of us alone.

Thoughts and rocks are intertwined in the unfolding of nature, as Michelangelo's *David* so eloquently attests. Yet is it possible to understand rationally how two completely different kinds of things can interact with each other? Logic says no, and history confirms that verdict. To form a rational comprehension of the interplay between the matterlike and mindlike parts of nature these two components ought to be understood as aspects of some single primal stuff. But what is the nature of a primal stuff that can have mind and matter as two of its aspects?

An answer to this age-old question has now been forced upon us. Physicists, probing ever deeper into the nature of matter, found that they were forced to bring into their theory the human observers and their thoughts. Moreover, the mathematical structure of the theory combines in a marvelous way the features of nature that go with the concepts of mind and matter. Although it is possible, in the face of this linkage, to try to maintain the traditional logical nonrelatedness of these two aspects of nature, that endeavor leads to great puzzles and mysteries. The more reasonable way, I believe, is to relinquish our old metaphysical stance, which though temporarily useful was logically untenable, and follow where the new mathematics leads.

This volume brings together several works of mine that aim to answer the question: How are conscious processes related to brain processes? My goal differs from that of most other quantum physicists who have written about the mind–brain problem. It is to explain how the content of each conscious human thought, as described in psychological terms, is related to corresponding processes occurring in a human brain, as described in the language of contemporary physical science. The work is based on a substantial amount of empirical data and a strictly enforced demand for logical coherence. I call the proposed solution the Heisenberg/James model because it unifies Werner Heisenberg's conception of matter with William James's idea of mind.

The introduction, "... and then a Miracle Occurs", was written specially for this volume. It is aimed at all readers, including workers in psychology, cognitive science, and philosophy of mind. Those fields, like physics, have witnessed tremendous changes during the century since William James wrote his monumental text. My introduction places the Heisenberg/James model in the context of that hundred-year development.

The main features of the model are described in "A Quantum Theory of the Mind-Brain Interface". This paper is an expanded version of a talk I gave at a 1990 conference, Consciousness Within Science. The conference was attended by neuroanatomists, neuropsychologists, philosophers of mind, and a broad spectrum of other scientists interested in consciousness. The talk was designed to be understandable by all of them, and the paper retains some of that character. Together with the introduction and appendix ("A Mathematical Model") it is the core of the present volume.

"The Copenhagen Interpretation" is an older paper of mine, reprinted from the *American Journal of Physics*. It describes the Copenhagen interpretation of quantum theory. That interpretation held sway in physics for six decades, and it represents our point of departure.

The other papers deal with closely related issues. Many of the ideas are to be found in my first published work on the problem, the 1982 paper "Mind, Matter, and Quantum Mechanics", from which this volume takes its title. An overview of the model is given in "A Quantum Theory of Consciousness", which summarizes a talk I gave at a 1989 conference on the mind–brain relationship.

The theory of the mind–brain connection described above is based on Heisenberg's ideas, and it accepts his position that the element of chance is to be regarded as primitive. Einstein objected to this feature of orthodox quantum thought, and Wolfgang Pauli eventually tried to go beyond the orthodox view, within the context of a psychophysical theory that rested in part on work of C. G. Jung. The possibility of extending the present theory in this way is discussed in "Mind, Matter, and Pauli".

"Choice and Meaning in the Quantum Universe" first describes some attempts by physicists to understand the nature of reality, and then attempts to discern, tentatively, a meaning intrinsic to natural process itself from an analysis of the form of that process alone, without tying meaning to any outside thing.

The mind–body problem is directly linked to man's image of himself, and hence to the question of values. The Heisenberg/James model of mind and man is separated by a huge logical gulf from the competing Cartesian model, which has dominated Western philosophic and scientific thought for three centuries. Two of the included papers, "Future Achievements to Be Gained through Science" and "A Quantum Conception of Man", were presented at international panels dealing with human issues, and they explore the potential societal impact of replacing the Cartesian model of man by the Heisenberg/James model. The second of these papers is the best introduction to this book for readers interested in seeing the bottom line before going into the technical details of how it is achieved.

The final chapter, "Quantum Theory and the Place of Mind in Nature", is a contribution to the book *Niels Bohr and Contemporary Philosophy*, which is to appear this year. It examines the question of the impact of quantum theory upon our idea of the place of mind in nature. This article can serve as a short philosophical introduction to the present volume, although it was a subsequent development in the evolution of my thinking.

In the above works I have tried to minimize the explicit use of mathematics. But in an appendix prepared for this volume I have transcribed some key features of the model from prose to equations.

Among the scientists and philosophers who have suggested a link between consciousness and quantum theory are Alfred North Whitehead, Erwin Schrödinger, John von Neumann, Eugene Wigner, David Albert and Barry Loewer, Euan Squires, Evans Harris Walker, C. Stuart, Y. Takahashi, and H. Umezawa, Amit Goswami, Avshalom Elitzur, Alexander Berezin, Roger Penrose, Michael Lockwood, and John Eccles. Only the final two authors address in any detail the problem addressed here: the nature of the relationship between the physical and physiological structures. Eccles's approach is fundamentally different from the present one. Lockwood's approach is more similar, but takes a different tack and does not attain the same ends.

Berkeley, February 1993   *Henry P. Stapp*

Acknowledgements

1 Supported by the Director, Office of Energy Research, Office of High Energy and Nuclear Physics, Division of High Energy Physics of the U.S. Department of Energy under Contract DE-AC03-76F00098.

2 Supported by the Director, Office of Energy Research, Office of High Energy and Nuclear Physics, Division of High Energy Physics of the U.S. Department of Energy under Contract DE-AC03-76F00098. Reprinted, with permission of the publishers, from *The Interrelationship between Mind and Matter*, edited by Beverley Rubic, Center for Frontier Sciences, Temple University, 1992.

3 Supported by the Director, Office of Energy Research, Office of High Energy and Nuclear Physics, Division of High Energy Physics of the U.S. Department of Energy under Contract DE-AC03-76F00098. Reprinted, with permission of the publishers, from the *American Journal of Physics* **40**, 1098–1116 (1972).

4 Supported by the Director, Office of Energy Research, Office of High Energy and Nuclear Physics, Division of High Energy Physics of the U.S. Department of Energy under Contract W-7405-EN-G-48. Reprinted, with permission of the publishers, from *Foundations of Physics* **12**, 363–399 (1982).

5 Supported by the Director, Office of Energy Research, Office of High Energy and Nuclear Physics, Division of High Energy Physics of the U.S. Department of Energy under Contract DE-AC03-76SF00098. Invited presentation to the conference "Consciousness Within Science", held at Cole Hall, University of California at San Francisco, 17–18 February 1990. Sponsored by the Bhaktivedanta Institute.

6 Supported by the Director, Office of Energy Research, Office of High Energy and Nuclear Physics, Division of High Energy Physics of the U.S. Department of Energy under Contract DE-AC03-76SF00098. Invited lecture at the Symposium on the Foundations of Modern Physics 1992:

"The Philosophic Thought of Wolfgang Pauli", held in Helsinki, 10–12 August 1992.

7   Supported by the Director, Office of Energy Research, Office of High Energy and Nuclear Physics, Division of High Energy Physics of the U.S. Department of Energy under Contract DE-AC03-76SF00098. Invited presentation at the congress "Science et Tradition; Perspectives Transdisciplinaires, Ouvertures vers le XXIème Siècle", UNESCO, 2–6 December 1991.

8   Contribution to the panel discussion "The Permanent Limitations of Science", sponsored by the Claremont Institute, Claremont, California, 14–16 February 1991. Other panelists: Roger D. Masters, Leon Kass, Edward Teller, Fred Hoyle, Stanley Jaki, Robert Jastrow.

9   Supported by the Director, Office of Energy Research, Office of High Energy and Nuclear Physics, Division of High Energy Physics of the U.S. Department of Energy under Contract DE-AC03-76SF00098. Invited paper for the Third UNESCO Science and Culture Forum—"Toward Eco-Ethics: Alternative Visions of Culture, Science, Technology, and Nature", held in Belem, Brazil, 5–10 April 1992. The introductory section of this paper was written in collaboration with Olivia B. Stapp.

10 Supported by the Director, Office of Energy Research, Office of High Energy and Nuclear Physics, Division of High Energy Physics of the U.S. Department of Energy under Contract DE-AC03-76SF00098. Contribution to the volume *Niels Bohr and Contemporary Philosophy*, edited by Jan Faye and Henry J. Folse (Kluwer, Dordrecht, 1993).

11 Based on a section from "The Volitional Influence of the Self and Mind (with Respect to Emotional Self-regulation)", by Jeffrey M. Schwartz, Henry P. Stapp, and Mario Beauregard in *Consciousness, Emotional Self-regulation and the Brain*, edited by Mario Beauregard (John Benjamins, Amsterdam & Philadelphia, 2003).

12 Based on a talk delivered at the "Future Visions" conference sponsored by the International Space Sciences Organization and the John Templeton Foundation that was held in conjunction with the annual State of the World Forum meeting 4–9 September 2000 in New York City.

13 Supported by the Director, Office of Science, Office of High Energy and Nuclear Physics, of the U.S. Department of Energy under Contract DE-AC02-05CH11231. I thank Ed Kelly for many useful suggestions pertaining to the form of this paper.

14 Supported by the Director, Office of Science, Office of High Energy and Nuclear Physics, of the U.S. Department of Energy under Contract

DE-AC02-05CH11231. I thank Efstratios Manousakis, Kathryn Laskey, Edward Kelly, Tim Eastman, Ken Augustyn, and Stan Klein for valuable suggestions.

Contents

Part I

Introduction

1 . . . and then a Miracle Occurs

A satisfactory understanding of the connection between mind and matter should answer the following questions: What sort of brain action corresponds to a conscious thought? How is the content of a thought related to the form of the corresponding brain action? How do conscious thoughts guide bodily actions?

Answers to these questions have been heretofore beyond the reach of science: the available empirical evidence has been unable to discriminate between alternative theories. Recently, however, mind/brain research has provided powerfully discriminating data that lift these questions from the realm of philosophy to that of science and lend strong support to definite answers.

In attempts to understand the mind–matter connection it is usually assumed that the idea of matter used in Newtonian mechanics can be applied to the internal workings of a brain. However, that venerable concept does not extrapolate from the domain of planets and falling apples to the realm of the subtle chemical processes occurring in the tissues of human brains. Indeed, the classical idea of matter is logically incompatible with the nature of various processes that are essential to the functioning of brains. To achieve logical coherence one must employ a framework that accommodates these crucial processes. A *quantum* framework must be used in principle.

Quantum theory is sometimes regarded as merely a theory of atomic phenomena. However, the peculiar form of quantum effects entails that ordinary classical ideas about the nature of the physical world are profoundly incorrect in ways that extend far beyond the properties of individual atoms. Indeed, the model of physical reality most widely accepted today among physicists, namely that of Heisenberg, has gross large-scale nonclassical effects. These, when combined with contemporary ideas about neural processing, lead to a simple model of the connection between mind and brain that is unlike anything previously imagined in science. This model accommodates the available empirical evidence, much of which is highly restrictive and from traditional viewpoints extremely puzzling.

Competing theories of the mind–brain connection seem always to have a logical gap, facetiously described as ". . . and then a miracle occurs". The model arising from Heisenberg's concept of matter has no miracles or special features beyond those inherent in Heisenberg's model of physical reality itself. The theory fixes the place in brain processing where consciousness enters, and explains both the content of the conscious thought and its causal efficacy.

This model of the mind/brain system is no isolated theoretical development. It is the rational outcome of a historical process that has occupied most of this century, and that links a series of revolutions in psychology and physics. Although the model can be discussed in relative isolation, it is best seen within the panorama of the twentieth-century scientific thought from which it arose.

The historical and logical setting for these developments is the elucidation by William James, at the end of nineteenth century, of the clash between the phenomenology of mind and the precepts of classical physics. I shall presently describe some of James's key points, and will then review, from the perspective they provide, some of the major twentieth-century developments in psychology: the behaviorist movement, the cognitive revolution, and the dominant contemporary theme, materialism. On the physics side, the crucial developments are Einstein's special theory of relativity, quantum theory, the Einstein–Podolsky–Rosen paradox, and the development of some models of physical reality that meet the demands imposed by the nature of quantum phenomena. Among these models the one proposed by Heisenberg is, in my opinion, the best. Coupled to James's conception of mind it produces a model of the mind–matter universe that realizes within contemporary physical theory the idea that brain processes are causally influenced by subjective conscious experience.

This model of the mind/brain links diverse strands of science, principally physics, psychology, and brain physiology. I shall endeavor to provide the necessary background in all three areas. However, I do not follow historical order but construct instead a rational narrative.

The first critical point, which underlies everything else, is the fact that the peculiarities of nature revealed by quantum phenomena cannot be dismissed as esoteric effects that appear only on the atomic scale. The Einstein–Podolsky–Rosen paradox, by itself, makes manifest the need for a radical restructuring of our fundamental ideas about the nature of physical reality. It also shows that this restructuring cannot be confined to the atomic scale. Quantum physicists have for years been proclaiming this need for a profound revision of ordinary ideas about the nature of the physical world. But their reasons have usually been based upon interpretations of atomic phenomena that are accessible only to experts in the field. To outsiders the whole business has remained shrouded in mystery. But the EPR paradox is a puzzle that can be expressed wholly in terms of behaviors of objects that are directly observable to the unaided eye.

To convince the reader that something is fundamentally wrong with ordinary ideas about nature I shall begin with a description of this paradox.

1.1 The Einstein–Podolsky–Rosen Paradox

In 1935 Albert Einstein, Boris Podolsky, and Nathan Rosen wrote a famous paper[1] that led to what is now seen to be an unexpected property of nature: an apparent need, at some deep level, for strong instantaneous actions over large distances. This conclusion, which is diametrically opposed to Einstein's own ideas about nature, is deduced from the predictions that quantum theory makes in certain special kinds of experimental circumstances. Typically, these are situations in which two experimenters perform at the same time, but in well-separated regions, independent measurements upon a single extended system. Each experimenter is allowed to freely choose—and then immediately perform—one of two alternative possible measurements on the large system. The combination of the two measurements, one performed by each of the two experimenters, is called here a *pair* of measurements.

In this situation there are four alternative pairs of measurements that might be performed. For each of these four pairs quantum theory makes an assertion about the connection between the outcomes of the two measurements. Einstein and his collaborators showed that these assertions, taken together, conflicted with strongly held ideas about the nature of physical reality. Over the years important generalizations of the original EPR arguments have been constructed, and the conflict has been sharpened considerably.

The most recent version of the EPR paradox is based on an experiment devised by Lucien Hardy.[2] The experimental details are unimportant in the present context. What *is* important is that a certain experimental procedure is used to produce a large collection of similarly prepared systems, and that each of these systems is then subjected to a *pair* of measurements. These two measurements are performed at the same time in two far-apart regions. The measurement performed in each region will be one of two alternative possible measurements, and the outcome of each performed measurement will be one of two alternative possible outcomes.

To make the description more pictorial, without changing the logic, I shall say that one of the two alternative measurements in each region

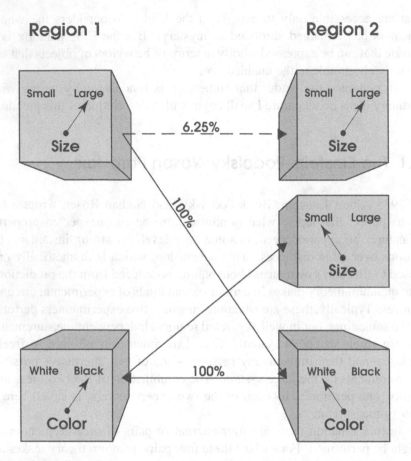

Figure 1   A diagrammatic representation of the predictions of quantum theory for the Hardy version of the EPR experiment.

measures "color" and the other measures "size". These two words are just a graphic shorthand for the two particular measurements that have been described in detail by Hardy. The device that measures "color" fills a one-cubic-foot box, and has a visible pointer that swings either to a position marked "black" or to a position marked "white". The device that measures "size" is a similar device, with positions marked "large" and "small". *One or the other* of the two possible measurements can be performed in each region, not both.

Quantum theory, transcribed into our language, makes four assertions pertaining to this situation. It will be shown that these four assertions, taken together, are logically incompatible with the following reasonable-sounding locality assumption: the last-minute choice by the experimenter in one region about which of the two measurements he will perform in that region cannot affect an outcome that appears far away at the same time under a fixed faraway experimental condition. This assumption is similar to the key locality assumption used by Einstein, Podolsky, and Rosen.

The four assertions of quantum theory are these (see Figure 1):

1 If "size" were to be measured in region 1 and the outcome there were to be "large", then if "color" were to be measured in region 2 the outcome there would be "white".

2 If "color" were to be measured in region 2 and the outcome there were to be "white", then if "color" were to be measured in region 1 the outcome there would be "black".

3 If "color" were to be measured in region 1 and the outcome there were to be "black", then if "size" were to measured in region 2 the outcome there would be "small".

4 If "size" were to be measured in both regions, then, in a large collection of paired measurements, both outcomes will be "large" in approximately one-sixteenth of the instances.

I shall now show how these four assertions of quantum theory, combined with our assumption of no action at a distance, lead to a logical contradiction. Readers not interested in following though the details of the logical argumentation can skip to the end of the section in small type.

The argument goes as follows. Suppose predictions 1 and 2 of quantum theory are correct. And suppose that no matter which of the two alternative possible measurements is performed in region 1 the outcome appearing there must be independent of which measurement is performed in region 2. And suppose a similar property with regions 1 and 2 interchanged also holds. Then the following conclusion holds:

Conclusion A. Suppose "size" is measured in region 1 and the outcome there is "large". Then if "color" instead of "size" had been measured in region 1, the outcome there would necessarily have been "black".

To verify this result suppose that "size" is measured in region 1 and that the outcome there is "large", just as the supposition of conclusion A demands. Suppose, moreover, that "color" is measured in region 2, just as the condition of prediction 1 of quantum theory demands. Then this prediction implies that the result in region 2 must be "white". Given this result "white", and the assertion that this outcome in region 2 cannot depend upon which measurement is performed in region 1, prediction 2 of quantum theory implies that if "color" instead of "size" had been measured in region 1, then the outcome there would necessarily have been "black".

This is the claimed conclusion. However, one extra assumption was used: it was assumed that "color" was measured in region 2. That condition can be dropped. For one of our assumptions is that no matter which measurement is performed in region 1 the outcome there must be independent of which measurement is performed in region 2. Hence the connection established between results in region 1 cannot be disturbed by changing what we do in region 2.

The natural interpretation of conclusion A is that, under the conditions of the experiment, whatever is measured in region 1 is "black" if it is "large". However, that inference goes beyond what is actually proved, for it depends on the additional assumption that there is an existing "something" that "has" the properties that are measured. But one idea in quantum theory is that there may be nothing in nature that possesses simultaneously the two properties that are represented here by the words "large" and "black". We do not wish to prejudge that idea, and hence will stick with our more conservative conclusion A.

Conclusion A combined with prediction 3 of quantum theory yields:

Conclusion B. Suppose "size" is measured in region 1 and the outcome appearing there is "large". Then if "size" is measured also in region 2, the outcome appearing in region 2 must be "small".

The assumption here is exactly the assumption of conclusion A. Hence we can use the conclusion: if "color" instead of "size" had been measured in region 1, then the outcome in region 1 would necessarily have been "black". But then prediction 3, coupled with the assertion that the result in region 2 cannot depend on which measurement is performed in region 1, implies that the result of the measurement of "size" in region 2 must be "small". This is what conclusion B asserts.

Conclusion B contradicts prediction 4 of quantum theory. Thus the predictions of quantum theory are logically incompatible with the assertion that the outcome of any measurement performed on one part of a quantum system must be independent of which measurement is performed simultaneously on a faraway part: large quantum systems seem to behave, at least in some special situations, as if they were instantaneously linked-up wholes.

The entire argument refers only to large visible objects, namely the macroscopic positions of devices and their pointers. Indeed, the predictions of orthodox quantum theory are, in principle, always assertions about such observable things. The details of the procedure by which these predictions are derived is not germane to our conclusion, which is simply that *the predictions themselves* are incompatible with the EPR assumption that no influence can act instantaneously over large distances.

EPR-type paradoxes are not just freak anomalies in quantum theory: they pervade the theory. Discovered in the mid-1930s by Schrödinger and Einstein, they have been, ever since, a chief focal point of the study of the foundations of physics. Numerous international conferences of physicists and philosophers have centered on the EPR problem, and references to "EPR" are ubiquitous in the foundational literature. The EPR-type phenomena apparently entail the need for strong instantaneous influences, at some deep level, and this evidently entails, in turn, the need for a major restructuring of our ideas about the fundamental nature of the physical universe.

Physicists have devised three alternative possible ways of understanding how the predictions of quantum theory can be valid. I shall describe these three models later. All are "radical": none conform to conventional ideas about the nature of the physical world.

I revert now to historical order.

1.2 James's Conception of Mind

James defines psychology as the science of mental life, where the latter includes such things as "feelings, desires, cognitions, reasonings, decisions, and the like".[3] He immediately distinguishes two possible ways of unifying the material, the spiritualistic, and the associationistic approaches. The former seeks to "affiliate the divers mental modes ... upon a simple entity, the personal soul", whereas the latter seeks "common elements *in* the divers mental facts rather than a common element behind them". In chapter I, after describing a host of disparate facts about mental life he says of the spiritualistic approach that

> our explanation becomes as complicated as the crude facts with which we started. Moreover there is something grotesque and irrational in the supposition that the soul is equipped with elementary powers of such an ingeniously intricate sort.[4]

On the other hand, he argues that

> the pure associationist's account of our mental life is almost as bewildering as that of the pure spiritualist. This multitude of ideas, existing absolutely, yet clinging together, and weaving an endless carpet of themselves, like dominoes in ceaseless change, or the bits of glass in a kaleidoscope,— whence do they get their fantastic laws of clinging, and why do they cling in just the ways they do?[5]

James, in his answer, cites numerous instances of evident mind–brain connection to support the conclusion that

> the spiritualist and the associationist must both be "cerebralists", to the extent of at least admitting that certain peculiarities in the way of working of their own favorite principles are explicable only by the fact that the brain laws are a codeterminant of the result.[6]

This conclusion elevates the problem of the mind–brain interaction into a place of central importance in Jamesian thought.

After an extensive review of habit and reflex action, James raises the issue of the automaton theory:

> The conception of reflex action is surely one of the best conquests of psychological theory; why not be radical with it? Why not say that just as the spinal cord is a machine with a few reflexes, so the hemispheres are a

machine with many, and that that is all the difference? The principle of continuity would press us to accept such a view.[7]

... so simple and attractive is this conception from the consistently physiological point of view, that it is wonderful to see how late it was stumbled on in philosophy, and how few people, even when it is explained to them, fully and easily realize its import.[8]

Descartes made a step in the direction of this "conscious automaton theory",

but it was not till 1870, I believe, that Mr. Hodgson made the decisive step, by saying that feelings, no matter how intense they may be present, can have no causal efficacy whatever.[9]

James goes on to recount hearing a most intelligent biologist say:

"It is high time for scientific men to protest against the recognition of any such thing as consciousness in a scientific investigation."[10]

James's rejoinder:

In a word, feeling constitutes the "unscientific" half of existence, and any one who enjoys calling himself a "scientist" will be only too happy to purchase an untrammeled homogeneity in terms of the studies of his predilection, at the slight cost of admitting a dualism which, in the same breath that it allows to mind an independent status of being, banishes it to a limbo of causal inertness, from which no intrusion or interruption on its part need ever be feared.[11]

James cites, nevertheless, one reason for accepting the causal inertness of consciousness:

Over and above this great postulate that things must be kept simple, there is, it must be confessed, still another highly abstract reason for denying causal efficacy to our feelings. We can form no positive image of the *modus operandi* of a volition or other thought affecting the cerebral molecules.[12]

He quotes, from an "exceedingly clever writer", a passage that ends with the sentences:

"Try to imagine the idea of a beefsteak binding two molecules together. It is impossible. Equally impossible is it to imagine a similar idea loosening the attractive forces between two molecules."[13]

This seeming impossibility of even imagining how an idea, or a thought, could influence the motions of molecules in the brain is certainly a main support for the highly counterintuitive notion that mind cannot influence matter. If there were a simple model showing how such an influence could occur, in a completely natural way, and within the framework of the established laws of physics, then the notion that our thoughts cannot effect our actions would undoubtedly lose much of its appeal.

James continues to quote the same author:

"Having firmly and tenaciously grasped these two notions, of the absolute separateness of mind and matter, and of the invariable concomitance of a mental change with a bodily change, the student will enter on the study of psychology with half his difficulties surmounted."[14]

James retorts:

Half his difficulties ignored, I should prefer to say. For this "concomitance" in the midst of "absolute separateness" is an utterly irrational notion. It is to my mind quite inconceivable that consciousness should have *nothing to do* with a business to which it so faithfully attends. And the question, "What has it to do?" is one that psychology has no right to "surmount", for it is her plain duty to consider it.[15]

James makes a positive argument for the efficacy of consciousness by considering "the particulars of the distribution of consciousness". He says that the study made throughout the rest of his book "will show that consciousness is at all times primarily *a selecting agency*". It is present when choices must be made between different possible courses of action. Such a distribution would be understandable if consciousness plays a role in making, or actualizing, these selections; otherwise this distribution makes no sense.

Beyond this crucial issue of the efficacy of consciousness, James's principal claim, at the fundamental level, is the *wholeness, or unity, of each conscious thought*. Each thought has components, but the whole is, he claims, more than than just a simple collection of its components. The component thoughts are experienced together in a particular way that makes the experienced whole an essentially new entity. It is these whole thoughts that are the proper fundamental elements of psychological theory, not some collection of "elementary components" out of which our thoughts are assumed to be formed by simple aggregation.

The object of every thought, then, is neither more nor less than all that the thought thinks, exactly as the thought thinks it, however complicated the matter, and however symbolic the manner of thinking may be.

... however complex the object may be, the thought of it is one undivided state of consciousness.[16]

An analogous property holds for the brain:

The facts of mental deafness and blindness, of auditory and optical aphasia, show us that the whole brain must act together if certain conscious thoughts are to occur. The consciousness, which is itself an integral thing not made of parts, "corresponds" to the entire activity of the brain, whatever that may be, at the moment.[17]

The main conclusion of the present work is that James's ideas about mind and its connection to brain accord beautifully with the contemporary laws of physics. But between the writings of James and this conclusion lie the monumental twentieth-century revolutions in science.

1.3 The Special Theory of Relativity

The special theory of relativity was announced by Einstein in 1905. It is pertinent here for two reasons. First, it caused an important shift in the generally accepted idea of the nature of science. The simple mechanical picture of the universe that had been developing so successfully during the preceding three centuries had beguiled scientists into believing that this simple idea of nature was an accurate image of the real thing. That classical picture involved, as Newton himself had specified, an absolute and homogeneous space, within which things changed in an absolute and homogeneous time. The Newtonian picture entailed the concept of a universal "now": a present instant of time defined unambiguously for every point in space. By overturning, in the minds of scientists, this intuitive idea of the instant "now", on the grounds that it could not be empirically tested, Einstein gave credence to the the idea that every concept in physics should be empirically testable. Einstein himself later strongly opposed that interpretation, as we shall see, but the idea lived on: the broad view that the task of science was to enlarge man's "understanding" of nature gave way, temporarily, to a "positivistic" attitude, which tended to shun, and even scorn, any component idea that was not directly testable empirically.

The rejection of the idea of the instantaneous "now" entailed also rejection of the Newtonian idea of instantaneous action at distance—for "instantaneous" lost all meaning. This led to the second pertinent consequence of the special theory of relativity, the idea that no influence originating at a point "A" could produce an effect at a point "B" before something traveling at the speed of light could reach B from A. This idea of no faster-than-light influence, like positivism, was also later to come into question, as we shall see.

1.4 The Behaviorist Movement

William James, and other nineteenth-century psychologists, took conscious-ness to be the core subject matter of psychology, and introspection a neces-sary tool for investigating it. He recognized that "introspection is difficult and fallible", and he apparently recognized that the problem of the con-nection of conscious process to brain process was irresolvable within the framework of the classical physics of his day. He foresaw, accordingly, important changes in physics. Others, less patient, embraced the radical solution: redefine psychology so as to exclude these difficulties. In 1913 John B. Watson launched the behaviorist movement with just such an aim:

> The time seems to have come when psychology must discard all reference to consciousness; when it no longer need delude itself into thinking that it is making mental states the object of observation.
>
> Its theoretical goal is the prediction and control of behavior. Introspection forms no essential part of its methods . . .[18]

Referring to the functionalist approach to psychology Watson says:

> One of the difficulties in the way of a consistent functional psychology is the parallelistic hypothesis. If the functionalist attempts to express his formulations in terms which make mental states really appear to function, to play some active role in the world of adjustments, he almost invariably lapses into terms that are connotative of interaction. When taxed with this he replies that it is more convenient to do so and that he does it to avoid the circumlocutions and clumsiness that are inherent in a thoroughgoing parallelism. As a matter of fact I believe that the functionalist thinks in terms of interaction and only resorts to parallelism to give expression to his views. I feel that *behaviorism* is the only consistent and logical functionalism. In it one avoids both the Scylla of parallelism and the Charybdis of interaction.[19]

This passage discloses the clouding of the thinking of a psychologist by the dogmas of classical physics: the idea of an active interplay between mind and matter was dismissed as not even in contention.

The behaviorists sought to explain human behavior in terms of certain relatively simple mechanisms, such as stimulus and response, habit for-mation, habit integration, and conditionings of various kinds. It is now generally agreed that the simple mechanisms identified by the behaviorists cannot adequately account for the full complexity of human behavior. It is, of course, a completely proper part of the scientific method to try simple ideas first. However, in the light of the tremendous complexity of the human brain, it seems now naive to expect that its operation could be fully reduced to things significantly simpler than consciousness. Rather, consciousness is a comparatively simple aspect of the complex brain process. The availabil-ity to us of these glimpses, however flawed and fallible, into the complex workings of the brain provides scientists with insights that can be exploited. These "seemings" are data to be explained, and the need to explain them constrains our theories.

1.5 Quantum Theory

Explaining consciousness in terms of quantum theory is no help to a person to whom quantum theory is a mystery. Since most scientists in the field of mind/brain research are not quantum physicists, I must, to make this work broadly useful, dispel the mystery of quantum theory. That is my intention.

Quantum theory is a statistical theory: it deals with probabilities. If a particle is in box, and we don't know where, but we do know that every possible location is equally likely, then we can imagine dividing the box into a huge number of little cubes of equal size, and assigning an equal probability to each one. If more information becomes available then the probabilities assigned to the various little cubes might be changed. For each such little cube we might also have probabilistic information about the velocity that the particle would have, if it were in that little cube. To represent this further information we could imagine defining little six-dimensional regions in position-and-velocity space and assigning a probability to each one. This collection of probabilities would define a "probability distribution" for the particle: it would specify, for each of these little regions in position-and-velocity space, a probability for the "particle and its velocity" to be in that little region. This probability distribution would, in general, change with the passage of time.

If there is a device that detects these particles, then one can define a distribution that is similar to the one just described, but that specifies not the probability for the particle to be in each little six-dimensional region, but rather the detection efficiency for that region: i.e., the probability (per unit time) that if the particle is in that little region then the detector will register a "detection event". By combining these two distributions one can compute the full probability (per unit time) for a detection event to occur under the condition specified by the initial probability distribution.

The description just given applies to the case of a classical particle. However, the same formula for the "probability of a detection event" holds also in the quantum case. The only differences are these: first, the evolution of the probability distribution during the passage of time is governed by a different equation of motion; second, the quantity that was interpreted in the classical calculation as the probability in a little region in position-and-velocity space can be negative. This second difference shows that interpretations of the two individual parts of the detection probability formula cannot be the same as they were in the classical case. However, it is only the whole formula that really counts anyway: only *it* can be compared directly to experiment. What goes on unobserved, and unobservable, at the atomic level is unimportant to the practical man of science.

This practical, or pragmatic, approach to quantum theory is called the Copenhagen interpretation. It avers that we scientists should be content with rules that allow us to compute all empirically verifiable relationships between our observations. This view claims that no "deeper understanding" is really a proper part of science. The key issue, however, is whether by seeking to "understand" what is happening unobserved we might be able to extend the scope of the theory to include relationships that formerly were not perceived to exist, or that seemed to lie beyond the reach of science. This was the issue raised by Einstein when he said:

> It is my opinion that the contemporary quantum theory . . . constitutes an optimum formulation of [certain] connections [but] . . . offers no useful point of departure for future developments.[20]

In this connection it is interesting to reflect upon a conversation between Einstein and Heisenberg, recounted by the latter.

1.6 Conversation between Einstein and Heisenberg

Early in 1926 Heisenberg described the new quantum theory at a symposium in Berlin attended by Einstein. Later, in private, Einstein objected to the feature that the atomic orbits were left out. For, he argued, the trajectories of electrons in cloud chambers can be observed, so it seems absurd to allow them there but not inside atoms. Heisenberg, citing the nonobservability of orbits inside atoms, pointed out that he was merely following the philosophy that Einstein himself had used. To this Einstein replied:

> Perhaps I did use such a philosophy earlier, and even wrote it, but it is nonsense all the same.[21]

Heisenberg was "astonished": Einstein had reversed himself on the idea with which he had revolutionized physics!

To find the probable cause of this astonishing reversal one need only look at what Einstein had done between the 1905 creation of special relativity and the 1925 creation of quantum theory. He created, in 1915, the general theory of relativity. That theory welded the absolute space and absolute time of Newton into an absolute spacetime. There are no observers or measurements. Rather there is an entire plenum of unobservable spacetime points bound together by differential equations. In his ten-year search for the general theory Einstein was driven, not by any effort to codify data, but rather by demands for rational coherence, and an abstract "principle of equivalence". He sent the finished work to his friend Max Born saying that no argument in its favor would be given, because once the theory was understood no such argument would be needed. The critical tests were carried out, and the predictions of the theory were confirmed.

The general theory of relativity, as an intellectual achievement, surpassed by far the special theory. The general theory also undermined the two general conclusions of the special theory mentioned earlier, namely the claim of the virtue in science of strict adherence to positivism, and of the absence in nature of a preferred instant "now". As regards this latter point, many solutions of the equations of general relativity do have a preferred sequence of instantaneous "nows". Furthermore, the universe we are living in has a global preferred rest frame, which defines instantaneous "nows" empirically. This frame has recently been empirically specified to within several parts per million.

Certain important gains in science have arisen from adherence to positivistic philosophy. But if Einstein's experience is a good guide, then the demand for rational coherence can be expected to carry us still further.

In spite of the reservations of Einstein and others, the Copenhagen view appeared to satisfy most quantum physicists during the first half of the century. This was surely due in part to widespread acceptance of the belief that it was impossible to comprehend what was going on behind the visible phenomena. However, during the 1950s three possible models of "what was actually happening in nature" were devised, and the third was due to Heisenberg himself.

1.7 Contemporary Models of Physical Reality

In 1952 David Bohm propounded a model of the physical world that explains the predictions of quantum theory in an essentially mechanical way.[22] One key assumption is the existence of a preferred rest frame. This frame defines "instantaneous nows", and it permits the introduction of an instantaneous action at a distance. The second key assumption is that the "probability distribution" appearing in quantum theory exists as a real thing in nature herself, rather than as merely a construct in the minds of scientists. In classical physics the probability distribution is merely a construct in human minds, but in all models of reality that conform to the demands imposed by quantum theory the probability distribution, or something very similar to it, exists in nature herself, outside the minds of men. The third key assumption in Bohm's model is the existence of a classical world of point particles (and/or classical fields). This classical world is a physical world of the same kind that is postulated in classical physics.

Given these assumptions Bohm was able to devise an instantaneous extra force that depends on the (objectively existing) probability distribution, and that "maintains" this probability distribution, in the following sense: for any given probability distribution, imagine an ensemble of classical worlds originally distributed in position and velocity so as to conform to this given probability distribution; then this imaginary ensemble will continue forever to conform to this evolving distribution, provided each of the various worlds in this ensemble evolves under the influence of Bohm's force.

How does Bohm's model explain the EPR paradox described above?

Suppose that the device that measures color is constructed so the outcome "black" is indicated by a swinging of the pointer on the device to the right, and the outcome "white" is indicated by a swinging of the pointer to the left. Suppose, similarly, that the device that measures "size" will indicate "large" or "small" by a swinging of its pointer to the right or left, respectively. Then after the measurement interaction has occurred the pertinent macroscopic pointer will be in one location or the other, either swung to the right or swung to the left. It will not be anywhere in between. The probability distribution will, therefore, be separable into two distinct branches, one corresponding to each of the two alternative possible outcomes of the measurement. These two branches will be confined to two different regions, in terms of the position of the pointer. These two regions must be well separated, on the scale of visibly detectable differences, if the two alternative possible outcomes are to be readily distinguishable by direct observation of the pointer.

But which of the two alternative possible outcomes "actually occurs"?

That is determined, in Bohm's model, by where the classically described pointer ends up, after the measurement operation has been completed. For example, the outcome is identified as being "black" if the classical pointer ends up in the swung-to-the-right region, or "white" if the classical pointer ends up in the swung-to-the-left region. This final position of the pointer will depend, of course, upon the forces that have been acting on the pointer. However, the forces needed to make the observable outcomes of the various alternative possible measurements conform to the assertions of quantum theory must be "nonlocal", in the following sense: the forces acting on objects in one region must, in some cases, depend upon *which experiment is performed in the other region*. But if one allows forces of this nonlocal kind then there is no problem in resolving the EPR paradox. For the paradox arises from a tacit assumption (made very explicit by Einstein, Podolsky, and Rosen) that what happens in one time and place must be independent of what an experimenter, acting at the same time in a faraway region, decides to measure. However, Bohm's force involves instantaneous action at a distance, and it leads to an explicit violation of this plausible-sounding no-faster-than-light-influence assumption, which is part of our ordinary classical idea about how nature operates.

In spite of this occurrence of faster-than-light influences, Bohm's model can reproduce all of the predictions of a relativistic quantum field theory. Moreover, it permits no faster-than-light *control* of events in one region by human decisions made in another. Thus the nonlocal character of Bohm's model of reality is *veiled*: the nonlocal character of the force, although explicit in the model, is inaccessible to us at the practical level.

The key features of Bohm's model are that the probability distribution *exists objectively*, and decomposes dynamically *at the level of the macroscopic variables*, during certain "measurement-type" physical processes, into distinguishable branches, one of which is singled out by a nonlocal mechanism. It is postulated that only this one singled-out branch is experienced in human consciousness.

The second kind of model of physical reality proposed by quantum physicists is one in which the probability distribution again exists objectively, and decomposes, just as before, into distinguishable branches at the macroscopic level. However, no mechanism selects one of these branches as the unique branch that is experienced in human consciousness: there exists, in the fullness of nature, a conscious experience corresponding to *each* of the alternative possible outcomes of each of the measurements. The fact that, for example, a certain pointer appears to any community of communicating observers to have swung only one way, or only the other way, not both ways at once, is understood in terms of the idea that the universe splits, at the macroscopic level, into various *noncommunicating branches*. I shall not endeavor to explain here how this works. But the fact that such an imaginative model is under serious consideration by mainline physicists indicates that it is a nontrivial task to devise a coherent model of the physical world that conforms, *even at the macroscopic level*, to the demands imposed upon models of physical reality by the nature of quantum phenomena.

The present work is based on Heisenberg's model of physical reality, or rather upon my elaboration of his model, which he did not describe in great detail. Heisenberg's model is simpler than either of the others. It dispenses with Bohm's classical physical world. However, it retains the idea that the probability distribution that occurs in quantum theory exists in nature herself. Indeed, in Heisenberg's model this probability distribution, and its abrupt changes, becomes the complete representation of physical reality. This shift from Bohm's manifestly dualistic representation of physical reality to a somewhat more homogeneous one is compensated, however, by a shift to a dualistic dynamics. The dynamical evolution of the physical world— as represented by this probability distribution—proceeds by an alternation between two phases: the gradual evolution via deterministic laws analogous to the laws of classical physics is *punctuated*, at certain times, by sudden uncontrolled quantum jumps, or events.

The essential features of the Heisenberg model of reality can be exhibited by considering again the EPR paradox. During the first phase of the measurement process the orderly evolution in accordance with the deterministic law of motion causes the probability distribution to develop in the same way as in Bohm's model: the probability associated with the macroscopic pointer position becomes concentrated in the two separated regions, where each region is associated with one of the two alternative possible outcomes of the measurement. Thus if a device that measures "color" is in place then the "probability" will become concentrated in two regions, one where the pointer on that device has swung to the right, to signify the outcome "black", and one where the pointer has swung to the left, to signify the outcome "white". In Bohm's model there is, in addition to this probability distribution, a real classical world, and the determination of which of the two alternative possible outcomes actually occurs is specified by whether the classical pointer ends up in the swung-to-the-right or the swung-to-the-left region. In Heisenberg's model there is no such classical world. Rather, it is postulated that *after the deterministic laws of motion have decomposed the probability distribution into the two well-separated branches* a "detection event" occurs. This event is a quantum jump, and it actualizes one or the other of these two alternative *macroscopic* possibilities, and eliminates the other. These Heisenberg events are considered to be the things that "actually occur" in nature: they are actual happenings, and they determine, by the selections they actualize, the course of physical events.

Heisenberg's model is structurally simpler than Bohm's because it does not involve plotting out the intricate motion of a classical world under influence of the nonlocal force. Also, once a Heisenberg detection event occurs the branch of the probability distribution that represents the *undetected* possibility is eradicated. In Bohm's model it awkwardly continues to exist.

The quantum jump actualizes, then, one or the other of the two macroscopic possibilities *previously generated by the deterministic laws of motion*. According to Heisenberg's idea the strength of the "tendencies" for the actualizations of the various alternative possibilities is specified by the (objectively existing) probability distribution itself. This ensures that the predictions of quantum theory will be satisfied.

Heisenberg's model of physical reality, as elucidated here, has three characteristics that are important in what follows: (1) the model postulates the existence in nature of "events", which are identified as the *actual happenings in nature*; (2) each such event actualizes a *large-scale happening*; it saves an entire macroscopic pattern of activity, and eradicates the alternatives; (3) such an event can occur only *after* an initial mechanical phase has constructed the distinct alternative macroscopic possibilities between which the choice is to be made.

The model of the mind/brain to be introduced here is based on the physical similarity between brains and measuring devices. Certain Heisenberg events that actualize large-scale patterns of neuronal activity in human brains will be identified as the physical correlates of human conscious events. The critical condition for such an identification is that the two correlated events (i.e., the physical event in the brain and the psychic event in the mental world) be images of each other under a mathematical isomorphism that is described in one of the papers that follow. This isomorphism maps conscious events in a psychological realm to corresponding Heisenberg events in a physicist's description of a brain.

To link this model of mind and brain to contemporary ideas in psychology I shall mention briefly the cognitive revolution in psychology and then examine some works of influential writers who have argued against dualism.

1.8 The Cognitive Revolution

The development in physics during the 1950s of models of what might be going on behind the visible macroscopic phenomena was matched a decade later by a parallel development in psychology. Advances in linguistics made it clear that the concepts identified by the behaviorists were too simple to account adequately for all of the complexities in human behavior. The examples of huge computing machines with complex software provided illustrations of how "cognitive", i.e., thoughtlike, processes can be generated in complex, albeit mechanical, ways by using internal *representations* of things external to the computer. These symbols for outside things can be created by the computer and interpreted by it. The brain could thus be imagined to be analogous to a computer, and the mathematics developed in connection with artificial intelligence imported into psychology. But the connection between cognition and consciousness was left unresolved by this development, and "mental" came to mean cognitive rather than conscious, because cognition was what could be dealt with. However, our concern is with consciousness.

1.9 Gilbert Ryle and Category Errors

Daniel C. Dennett is an influential author, and philosopher of mind. His book *Consciousness Explained* has a section entitled "Why Dualism Is Forlorn". It begins with the words:

> The idea of mind as distinct ... from the brain, composed not of ordinary matter but of some other, special kind of stuff, is *dualism*, and it is deservedly in disrepute today ... Ever since Gilbert Ryle's classic attack on what he called Descartes's "dogma of the ghost in the machine", dualists have been on the defensive. The prevailing wisdom, variously expressed and argued for, is *materialism*: there is one sort of stuff, namely *matter*—the physical stuff of physics, chemistry, and physiology—and the mind is somehow nothing but a physical phenomenon. In short, the mind is the brain.[23]

Bernard Baars, in his book *A Cognitive Theory of Consciousness*, also cites Ryle:

> ... philosopher Gilbert Ryle presented very influential arguments against inferred mental entities, which he ridiculed as "ghosts in the machine" and "humunculi". Ryle believed that all mentalistic inferences involved a mixing of incompatible categories, and that their use led to infinite regress.[24]

Because Ryle's 1949 arguments are still influential it is incumbent upon us to see how his proofs impact upon our model. The first preliminary step is to distinguish between two different kinds of mind: ghost-in-the-machine mind, and Jamesian mind.

James, at the end of his long chapter entitled "The Consciousness of Self", gives his conclusions:

> The consciousness of Self involves a stream of thought, each part of which as "I" can (1) remember those that went before, and know the things they knew; and (2) emphasize and care paramountly for certain ones among them as "*me*", and *appropriate to these* the rest ... This *me* is an empirical aggregate of things objectively known. The *I* that knows them cannot itself be an aggregate. Neither for psychological purposes need it be considered to be an unchanging metaphysical entity like the Soul, or a principle like the pure Ego, viewed as "out of time". It is a *Thought*, at each moment different from that of the last moment, but *appropriative* of the latter, together with all that the latter called its own ... *thought is itself the thinker*, and psychology need not look beyond.[25]

It is this "Jamesian mind" that our quantum model explains: it involves no "knower" that stands behind the thoughts themselves. Hence it is less susceptible to infinite regress.

Ryle gives several infinite-regress arguments. The first deals with intelligence and knowing. He distinguishes between intelligent behavior and the operation of thinking about what one is doing.

> This point is commonly expressed in the vernacular by saying that an action exhibits intelligence, if, and only if, the agent is thinking what he is doing while he is doing it, and thinking what he is doing in such a manner that he would not do the action as well if he were not thinking what he is doing . . . I shall argue that the intellectualist legend is false and that when we describe a performance as intelligent, this does not entail the double operation of considering and executing . . . The crucial objection to the intellectualist legend is this. The consideration of propositions is itself an operation the execution of which can be more or less intelligent, less or more stupid. But if for any operation to be intelligently executed a prior theoretical operation had first to be performed intelligently, it would be logically impossible for anyone to break into the circle . . . The regress is infinite, and this reduces to absurdity the theory that for an operation to be intelligent it must be steered by a prior intellectual operation. What distinguishes sensible from silly operations is not their parentage but their procedure, and this holds no less for intellectual than for practical performances . . . "thinking what one is doing" does not connote both "thinking what to do and doing it". When I do something intelligently, i.e. thinking what I am doing, I am doing one thing not two. My performance has a special procedure or manner, not special antecedents.[26]

In general, according to this view, mind refers to *a way a body can behave*, for example intelligently, not to some *thing* that belongs to the same category as a body.

Ryle's argument does not confute our dualistic model. For in this model the thinking and the doing do not occur in tandem. The thought and the physical act that implements it are two faces of a single mind/brain event. "Thinking what one is doing while one is doing it" is just that: the thought and doing are two aspects of a single event; hence they do not occur in tandem. Heisenberg's conception of physical reality leads to a mind/brain action, which, by combining the intellectual and functional aspects of the executive act into a single event, evades Ryle's attack on dualism.

Only the first of Ryle's infinite-regress arguments has been dealt with here, but all of those arguments fail for similar reasons to cover to the Heisenberg/James model. The essential point is that Ryle's arguments are directed against ghost-in-the-machine mind: they do not carry over to the Jamesian type of mind that occurs in the H/J model, in which the thought is the thinker is the feel of the actual brain event.

1.10 Dennett's *Consciousness Explained*

Daniel Dennett's book *Consciousness Explained* approaches the problem of consciousness from the materialist point of view. He announces that

> it is one of the main burdens of this work to explain consciousness without ever giving in to the siren song of dualism. What, then, is so wrong with dualism? Why is it in such disfavor?[27]

His answer cites the problem of understanding how mind can interact with matter:

> A fundamental principle of physics is that any change in the trajectory of a particle is an acceleration requiring the expenditure of energy...this principle of conservation of energy...is apparently violated by dualism. This confrontation between quite standard physics and dualism has been endlessly discussed since Descartes's own day, and is widely regarded as the inescapable flaw of dualism.[28]

This objection does not apply to the Heisenberg/James model. This model makes consciousness causally effective, yet it is fully compatible with all known laws of physics, including the law of conservation of energy.

Dennett adopts "the apparently dogmatic rule that dualism is to be avoided *at all costs*". He thus strips mind away from Cartesian dualism and arrives at a notion that plays a key role in his arguments: Cartesian materialism. This is the idea that there is

> a central [but material] Theater where "it all comes together"...a place where the order of arrival equals the order of "presentation" in experience because *what happens there* is what you are conscious of.[29]

Later he speaks of the audience, or witness, to presentations in this Cartesian Theater; or of the Ego, or Central Executive, or Central Meaner as the witness to such a presentation. These references to a "witness" seem to bring "mind" back in. But what Dennett wishes to confute, in order to buttress his own counterproposal, is the stripped-down idea of a "presentation" in a central Cartesian Theater, regardless of who, if anyone, is watching it.

Dennett's chief argument against the idea of a presentation in the Cartesian Theater is based on some experiments by Kolers and von Grünau: "Two different colored spots [separated by, say, 4 degrees] were lit for 150 msec each (with a 50 msec interval); the first spot seemed to begin moving and then change color abruptly *in the middle of its illusory passage* toward the second location." The puzzle is: "How are we able...to fill in the spot at the intervening place-times along a path running from the first to the second flash *before that second flash occurs*." This timing inversion is difficult to reconcile to the Cartesian Theater model of mind and brain. For within that model the information required to put the show on apparently arrives at the Theater only at the end of the show.

Dennett uses this difficulty with the Cartesian Theater model to justify his own approach, which he calls the Multiple Drafts model. That model rejects the intuitive idea of a single stream of consciousness.

> Instead of such a single stream (however wide), there are multiple channels in which specialist circuits try, in parallel pandemoniums, to do their various things, creating Multiple Drafts as they go.[30]

This claim that the stream of consciousness that "seems to exist" does not really exist is the crux of Dennett's theory. Responding to this surprising claim his fictional interlocutor, "Otto", exclaims:

> It seems to me that you've denied the existence of the most indubitably real phenomena there are: the real seemings that even Descartes in his *Meditations* couldn't doubt.[31]

Dennett replies:

> In a sense, you're right; that's what I'm denying exist.[32]

He elaborates by referring to a certain optical illusion in which there "seems to be a pink ring" even though there is no such ring in the external object being viewed. He asserts that

> there is no such thing as a pink ring that merely seems to be.[33]

> *There seems to be phenomenology* . . . But it does not follow from this undeniable, universally attested fact that there *really is* phenomenology.[34]

Dennett denies that experience is what it seems to be. He needs a strong argument to support such a counterintuitive claim. His argument is that the failure of the Cartesian Theater model rules out the stream of consciousness:

> There is no single, definitive "stream of consciousness", because there is no central Headquarters, no Cartesian Theater where "it all comes together" for a Central Meaner.[35]

That argument is not logically sound: the absence of a central place where order of arrival equals order of presentation does not logically entail that there can be no stream of complex unified thoughts of the kind we seem to have.

The fundamental problem here is how *can* one logically form entities that are intrinsically—*i.e., strictly within themselves, without the help of some outside binding agent*—complex wholes, within a logical framework that is fundamentally reductionistic—i.e., within a framework in which everything is asserted to be nothing but an aggregation of simple parts. Such a feat is a logical impossibility, and that is why James despaired of resolving the problem of mind within the framework of classical physics. In order to accommodate an intrinsically unified thought, as distinct from an aggregation that is *interpreted* as an entity by something else, one must employ a logical framework that is not strictly reductionistic: a framework that has among its logical components some entity or operation that forms wholes. A Heisenberg event is just such an element, and the Heisenberg/James model provides an explicit counterexample to Dennett's claim. This model has no Cartesian Theater, but it accommodates a stream of consciousness of the kind described by James. The empirical evidence that undermined the possibility of a Cartesian Theater is easily accommodated in the H/J model, so Dennett's line of argument is refuted by counterexample. In particular, the Kolers–von Grünau result is easily explained, as are all of the other "puzzling" experimental results he cites, such as Grey Walter's precognitive carousel, Geldard and Sherrick's cutaneous rabbit, and Libet's subjective delay.

Let me explain. First, the rudiments of brain dynamics must be understood. In a normal computing machine the currently active information is stored in a generally small number of registers. But in the brain a huge number of separate patterns of neural excitations can be present at one time. These patterns can become correlated to stimuli and responses, and can mediate the behavior of the organism. In a manner discussed in some detail in one of the following papers, the structure of these neural patterns can form *representations* of the body and its environment, with a history of the occurring representations becoming stored in memory. The main postulate of the model is that every conscious event is the psychological counterpart of a certain special kind of Heisenberg event in the brain, namely an event that actualizes a pattern of neuronal activity that constitutes a *representation* of this general kind. However, any such representation must be *formed before it can be selected*: the representation must be constructed by unconscious brain activity, governed by the preceding mechanical phase of the dynamical evolution, before it can be actualized. During this preliminary mechanical phase a superposition of many such representations must inevitably be generated. During the subsequent actualization phase *one* of these representations will be selected.

This general picture, applied to the Kolers–von Grünau case, means that the massive unconscious parallel processes of the brain will strive to construct a coherent picture of the changing environment, compatible with the available clues, *before* it is presented for possible adoption by a conscious event. In any good organization the executive action functions in this same way. In this process of constructing a coherent representation of the environment there is no central place in the brain where order of arrival equals order of "presentation" in consciousness. The representations of the evolving environment are constructed by fitting patterns of neuronal activity together in ways that conform to the rules for forming coherent representations of the evolving environment, but constrained by incoming data. The final patterns are essentially global, relative to the brain. A conscious event actualizes only a fully constructed and coherent representation of "the evolving self-and-surroundings", *after* it has has been formed by unconscious processes. The Kolers–von Grünau result is simply an instance of this general mode of mind/brain functioning. It arises naturally from this dualistic model that (1) is compatible with the laws of physics, (2) makes consciousness functional, and (3) identifies each conscious event, as described by James, as the image in a psychological realm of a special kind of Heisenberg event in the brain.

To understand more fully the character of these Heisenberg events consider a materialist's picture of a man in a "black box", isolated from all outside influences and observers. If this system is to be described in terms of strict laws of physics, and no observation or detection event occurs, then the system will evolve in accordance with the purely mechanical aspect of the law of motion. The parallel processors in the brain will churn out their various determinations, and the system will evolve, just like Schrödinger's notorious cat, into a superposition of macroscopically distinguishable systems. Whereas that famous cat developed into a superposition of an "alive cat" and a "dead cat", so the man-in-the-box will develop, under the action of the purely mechanical laws, into a superposition of, for example, a "standing man" and a "sitting man". However, if the man then observes himself, and a *detection event* occurs, then one of the two alternative possibilities will be selected, and the other will disappear from the realm of possibilities. This "observation" is not necessarily visual: "seeing" is not singled out as the unique way of fixing the macroscopic facts. The man could, instead, merely *feel* himself to be standing, or to be sitting: that sort of sensing also constitutes an "observation".

The H/J model postulates that this *sensing event* can be pushed back into the brain to the point where the brain's representation of the "standing man", or of the "sitting man", is fully formed. The sensing event then actualizes, in the brain, one or the other of these two representations. The associated experience is an image of this representation. This isomorphic connection, which is described in detail, is the core of the model.

This Heisenberg event actualizes precisely the sort of entity that is needed to guide effective action. Indeed, according to James's ideomotor theory, it is just this sort of sensed representation of bodily position and action, projected into the future, that acts as the mind/brain's template for bodily action. The book by Baars, *A Cognitive Theory of Consciousness*, presents contemporary empirical support for James's theory. Thus this second, or actualizing, phase of the dynamical process goes beyond the mechanical parallel-processing phase. It fixes a structure in the brain that, in the case of simple attention, is the brain's representation of the self and its surroundings, and, in the case of an intention to act, is the representation that serves as the template for the chosen course of action.

Another piece of empirical evidence cited by Dennett as contrary to the Cartesian Theater model, and hence evidence against the *actual* occurrence of the phenomenal stream of conscious events that *seems* to occur, is Grey Walter's precognitive carousel. In this experiment the subject views a sequence of slide projections, and is told that he can, when he wishes, advance to the next slide by pressing a button. But what actually advances the slide is an amplified signal from an electrode implanted into the subject's (patient's) motor cortex. The subjects "reported that just as they were 'about to' push the button, but before they had actually decided to do so, the projector would advance the slide".

This phenomenon is completely in line with the H/J model, which specifies that voluntary bodily actions are initiated by the actualization of a *projected* representation—a representation of what the forthcoming representations of the body and its environment are expected to be. Because the neural signal that is supposed to inaugurate the proceedings is intercepted, and the intended action initiated prematurely, the monitored representations of what actually takes place will fail to match this expectation. This mismatch will be experienced as a temporal anomaly in the subject's representation of "Self and Surroundings". The subject will be surprised at the "premature" motion of the slide, which he is, of course, unable to veto.

How different is the H/J model from Dennett's? The former, on the one hand, is far more specific about the connection between brain events and conscious events. It accepts the Jamesian stream of consciousness as raw data, and tries, with apparent success, to explain both the form of this data and its connection to brain processes. Dennett's theory, on the other hand, rather than explaining the data, seeks to circumvent it. By challenging the existence of the phenomenology, on the theoretical ground that there can be no Cartesian Theater, he tries to discredit all introspective data that is not backed up by objective data. However, Dennett's theoretical argumentation, like Ryle's, is directed essentially against Cartesian mind, rather than against Jamesian mind. Once the idea of "the witness" is replaced by the idea that each unified conscious experience simply *occurs*, in conjunction with a complex global brain event that constitutes a complex "judgement", one arrives at a "dualism" that may not be irreconcilably different from Dennett's "materialism". The latter does admit the existence of consciousness, and hence, to be complete, it must eventually describe how the relatively simple contents of conscious thoughts are related to the exceedingly complex processes that are occurring in the brain. Dennett will have to embrace quantum theory if he is to have a fully coherent model of the brain. Thus his "materialistic" model, with these details adequately filled in, could evolve into the H/J model.

The remnants of positivism have survived better in psychology than in other branches of science. Yet the merit of a scientific theory lies not in the verifiability of its individual parts. It lies in its internal consistency and economy, its scope and adequacy, and its cohesiveness with the rest of science. The Heisenberg/James model, though still in it infancy, does well on all these counts.

1.11 Comparison with Penrose

Two other physicists have written books propounding ideas related to those developed here. The central theme of both works is that the emergence of consciousness in association with brain processes is closely tied up with the quantum character of physical reality. Since that thesis is the core also of the works collected here, a comparison of my works with theirs is in order.

The first of the two books is Roger Penrose's *The Emperor's New Mind*. It describes quantum theory in some detail, focusing on the two very different ways in which a quantum system can evolve. The first is by the smooth deterministic development in accordance with the basic quantum law of motion, the Schrödinger equation. The second is by the sudden and unpredictable quantum jumps. In Penrose's terminology, the smooth development is called the "unitary" process U, and the abrupt one is called the "collapse" or "reduction" process R. Each quantum jump effects a "choice" or "decision" that picks out and actualizes *one* of the many "linearly superposed possibilities" previously generated by the unitary process U.

Near the end of his book Penrose arrives at the main conclusion:

> I am speculating that the action of conscious thinking is very much tied up with the resolving out of alternatives that were previously in linear superposition. This is all concerned with the unknown physics that governs the borderline between U and R . . .[36]

Penrose's book can be viewed as a detailed attempt to justify this idea.

If Penrose's conclusion is correct then several questions immediately arise: How are the structural features of our conscious thoughts related to the structural features of what is going on in the brain? Can conscious thoughts direct the course of brain activity? If so, how does a conscious thought produce, in some natural and understandable way, without invoking any magical or mystical transcription process, precisely the brain activity that promotes the goal represented in the thought? Penrose makes no attempt to answer these questions. They are, however, precisely the ones that I address. Thus my works begin where Penrose's book leaves off.

One of Penrose's main reasons for believing that the quantum character of reality is essential to the occurrence of consciousness is the shared "global" character of conscious thoughts and quantum states. I, also, have emphasized this point, but it may be useful to reinforce it by using Penrose's words.

Penrose cites many examples from the worlds of mathematics and music in which a thought seems to grasp an entire complex whole. One example is Mozart's description of how he creates a musical composition:

> "I keep expanding it, conceiving it more and more clearly until I have the entire composition finished in my head though it may be long. Then my mind seizes it as a glance of my eye a beautiful picture or a handsome youth. It does not come to me successively, with various parts worked out in detail, as they will later on, but in its entirety that my imagination lets me hear it."[37]

Penrose goes on to say that

> it seems to me that this accords with a putting-up/shooting-down scheme of things. The putting-up seems to be unconscious ("I have nothing to do with it") though, no doubt, highly selective, while the shooting-down is the conscious arbiter of taste ("those which please me I keep . . ."). The globality of inspirational thought is particularly remarkable in Mozart's quotation ("It does not come to me successively . . . but in its entirety"). Moreover, I would maintain that a remarkable globality is already present in our conscious thinking generally.[38]

Penrose cites a number of similar occurrences in the lives of Poincaré, Hadamard, and himself: after appropriate preparation the answer pops into consciousness as a complete unit.

To physicists who have long wrestled with the fundamental questions in quantum theory this "globality" of conscious thoughts is reminiscent of the "globality" of quantum states. Although this latter property is essentially technical it is worthwhile trying to convey its essence.

The "globality" of quantum states is revealed most strikingly in the Einstein–Podolsky–Rosen paradox mentioned earlier. Penrose correctly stresses that this paradox is *not* similar to the correlations between distant events that are understandable in the framework of ordinary classical ideas about the nature of the physical world.[39] In that classical way of thinking you can contemplate a situation in which two balls are shot out in opposite directions from a central region, and in which you know that one of the balls is white and the other is black. If you then find out that the ball appearing in your vicinity is white you can immediately infer that the ball appearing in the other vicinity must be black. There is no puzzle or paradox in any of this.

The EPR situation is quite different. There the combined system of two far-apart particles acts as a single global entity, in the sense that it *is not possible* to impose the following causality requirement: "*What a scientist decides to do* to one part of a system cannot affect in any way how the system will respond at the same instant to a measurement performed upon it far away." Thus a quantum system seems able to behave as a unified entity: What *you do to it* in one place can influence how it will react to a *simultaneous* probing far away.

The profoundness and irrevocability of this collapse of the classical local-reductionistic conception of the physical universe was not fully recognized even by Einstein, Podolsky, and Rosen: it became clear only after John Bell had prepared the way with his famous "Bell's theorem". The basic message of Penrose's book, as of mine, is that the enormous changes that have been wrought by quantum theory in our ideas about the fundamental nature of matter have altered radically the problem of the connection of mind to matter.

Penrose begins his book with a look at the popular question "Can a computing machine think?" Back in 1950 Alan Turing proposed that this question be replaced by a substitute, which is essentially this: "Can a computing machine *behave as if it thinks*, in the sense that it can normally answer questions from a human interlocutor well enough to fool that interlocutor into believing that the answers might be coming from a human being?"

The substitute question appears on the face of it to be inequivalent to the original: *thinking* is not idential to *behaving as if thinking*. But Turing believed that the original question was "too meaningless to deserve discussion". His principal point was that if the original question were converted into the less ambiguous form "Is this machine *conscious*, i.e., 'aware'?", then the issue is untestable. For the only way one can be sure about whether a machine is conscious is to *be* that machine! Consequently, the demand for intersubjective agreement, which is deemed essential to science, cannot be met: only *one* machine can actually *be* the given machine.

Turing's proposal was influential during the reign of behaviorism, but the fault with this kind of approach is now apparent: it limits theoretical creativity too severely. The concept of consciousness *might* be useful in the construction of a theory that economically, adequately, and comprehensively organizes the many kinds of empirical data pertaining to human behavior.

Hence the concept of consciousness ought not be summarily banned from science by a philosophical prejudice.

Penrose goes even further. He says that when he speaks of such things as thinking, feeling, understanding, or above all *consciousness*,

> I take the concepts to mean actual objective "things" whose presence or absence we are trying to ascertain, not to be merely conveniences of language.[40]

He says that he has in mind

> that at some time in the future a successful theory of consciousness might be developed—successful in the sense that it is a coherent and appropriate physical theory, consistent in a beautiful way with the rest of physical theory, and such that its predictions correlate precisely with human beings' claims as to when, whether, and to what degree they themselves seem to be conscious—and that this theory might indeed have implications regarding the putative consciousness of our computer.[41]

In the margin in my copy of Penrose's book, next to this passage, I have the annotation "Bravo!": Penrose and I are in close agreement on this point.

A good portion of Penrose's book is spent developing two arguments that I believe to be flawed, or at least inconclusive. The first argument attempts to justify the idea that the quantum reduction process **R** mentioned earlier—and hence also consciousness—is closely connected to *gravity*. The other argument is intended to support the thesis that human consciousness must access in some direct way the Platonic realm of abstract mathematical truth.

The first of these arguments is based on the assumption that if black holes exist, and if phase space disappears into them, then some other process must create compensating amounts of phase space in regions devoid of black holes. This is an awkward assumption that I find artificial and uncompelling. Even if it were accepted I doubt that the quantum process **R** would correctly achieve the end that it demands. For the process **R** *casts out* some of the classical branches of phase space populated by the flow generated by the unitary **U**. Hence **R** seems, if anything, effectively to eliminate phase space rather than create it.

In the papers assembled here I adhere to the more conventional idea that the process **R** is not connected in a special way to either gravity or the associated Planck mass of 10^{-5} grams, but that rather the quantum jumps effect choices between possibilities that are distinguishable at the level of directly observable phenomena. The deep questions *why* the jumps occur at this level, and *which* of the allowed possibilities is actually chosen in any individual quantum event, are not addressed here: the present form of the proposed quantum theory of the mind/brain interface—like the present orthodox quantum theory of matter—does not seek to answer either of these deep and still unresolved questions.

Penrose's second argument is intended to buttress his belief that human consciousness directly accesses the Platonic realm of abstract mathematical truth. Within the framework of contemporary scientific ideas this proposal must be classified as a "mystical" notion. I adhere instead to a "naturalistic" position that restricts theoretical entities to the union of the physical world of particles and fields, as it is described by quantum theorists, and the experienced world of thoughts and feelings, as it is described by psychologists.

Penrose's argument in favor of direct conscious access to the Platonic realm of abstract truth proceeds essentially as follows: he argues that since mathematicians can come to agreement about the truth or falsity of mathematical statements either they must all be appealing ultimately to the same "algorithm" (i.e., finite set of step-by-step rules) for determining mathematical truth, or they must sometimes use nonalgorithmic methods. The first possibility is ruled out because either the algorithm is known to the mathematicians or it is not, and if it were known to them then they could, by using the celebrated construction devised by Kurt Gödel, be able to prove the truth of a certain proposition that their algorithm cannot validate, which would mean that this algorithm cannot be their final arbiter of mathematical truth; but if the algorithm were not known to them then they would be accepting the verdicts of an algorithm whose validity they cannot have established, and this contradicts the mathematicians' conviction that they can be certain that the truths which they prove to be true must indeed be true. This leaves open only the second possibility, which is that nonalgorithmic methods must sometimes be used. But what is the nature of this nonalgorithmic way of divining the truth of mathematical statements?

If the physical universe indeed behaves in accordance with rules of quantum theory, then it acts in some sense algorithmically, provided the quantum element of chance is simulated by a (pseudo) random number generator. Thus there would be, in principle, at least in a closed quantum universe containing only a finite number of particles, and in some discrete step-by-step approximation, a universal algorithm controlling all brain processes, and hence, in our "naturalistic" theory, all conscious processes as well. Of course, our conscious thoughts contain representations of only a tiny part of what would be represented in a complete quantum-mechanical description of the brain. Thus it is at most only the general principles governing the activities of our mind/brain, not the full detailed description, that can ever be actually known to us. Yet from these general principles, expressed in terms of comprehensible mathematical rules, we ought to be able to understand quite well how the parts of our brain processes that correspond to certain "mathematical" portions of our thought processes can conform to logical rules, and hence produce correct algorithmic reasoning. However, the mathematical algorithms produced in this way are only a minor by-product of the general quantum algorithm that is controlling our mind/brain process, and they consist not of one single universal algorithm, but rather of an expanding collection of algorithms, which grows as mathematicians create more and more of these things. The question at issue is therefore this: How can we, at any stage in the development of mathematics, validate the truth of a mathematical statement whose truth is not implied by the rules that we have already validated?

This problem is to some extent a bootstrap one of demonstrating consistency: i.e., of proving the absence of any possibility of contraction. Yet, on the other hand, the system of mathematical truths consisting of true statements about the unending sequence of integers $1, 2, 3, \ldots$ seems to be more than just a matter of human creative prowess. We "know" that the properties of this set that we have proved to be true are indeed true. But *how* do we know it?

There does indeed seem to be some sort of "insight" or "intuition" involved here. But do we therefore really have to admit to divination via non-naturalistic access to a nonphysical realm of abstract mathematical truth? Might not "insight" be understood in terms of the quantum-mechanically described brain process?

Consider a simple example: What is the general formula for the sum $1 + 2 + 3 + \cdots + (N - 1) + N$? One way to arrive at the answer is to imagine a big square checker board that has N little squares along each side. Then imagine coloring in the bottom little square in the first (i.e., leftmost) column, the bottom two squares in the second column, the bottom three squares in the third column, $\ldots$, and finally the N little squares in the Nth column. In all we will have colored in the entire lower-right-half triangular portion of the big square, plus the upper-left-half triangular portions of each of the N little squares on the diagonal, for a total of $(N^2/2) + N/2 = N(N + 1)/2$ little squares. The total number of little squares filled in is the desired sum.

This process in our imagination allows us to "know" that the sum in question is $N(N + 1)/2$, no matter how big the number N is, without having previously formulated in any precise way the rules about how we actually know this to be true.

In this example it is not clear that we have any need to access any nonphysical realm of absolute truth. The argument seems to arise from the exercise of certain powers of manipulation and abstraction of visual images, and such powers might conceivably develop in a natural physical way from interactions, since birth, of our brains and bodies with our actual physical environment. This possibility seems to mesh well with the assertion by

Einstein cited by Penrose that in his (Einstein's) case the psychical entities of abstract thought are "of visual and some muscular type".[42]

Penrose's argument fails to really close off the possibility of this kind of physical explanation of mathematical insight. The reasoning process, like all mental processes, might be controlled by the universal quantum-mechanical rules, and hence be "mechanical" and (in some approximation) "algorithmic" and "naturalistic" although following rules unknown to the mathematician himself. The problem, then, is the origin of the mathematician's sense of certainty that certain conclusions produced by this process "must necessarily be true".

The answer might conceivably be that the mathematician, through an accumulation of experience involving cross-checking and continual testing of ways of manipulating images originating ultimately in common sensations, comes to have an exceedingly secure feeling about the "correctness" of certain procedures—correctness in the sense that these procedures always do lead to conclusions that hang together in a coherent and consistent way. The feeling of certainty could arise from the fact that the consistency of the way that a certain imagined sequence of visual images hangs together is "visually" obvious in the simplest instances, and it becomes apparent through experience that one can "increase N by one" without disrupting the relationships that were obvious in the simplest case. This explanation would demand certain powers to manipulate imagined visual images, and to discriminate which sequences hang together in a consistent way, but such powers with respect to visual imagery could be expected to arise from the interaction of the members of our species with the world about them, over an evolutionary epoch. There is no clear need, in the development of such powers, for access to some nonphysical realm of abstract mathematical truth.

This account of the origin of mathematicians' sense of certainly is not conclusively ruled out by Penrose's argument. In fact, this possibility is not actually addressed by that argument: it slips through the net of his logic because the reasoning process is in the physical sense "algorithmic" being based on the universal quantum rules, but in the mental sense "insightful" being based on the mathematician's almost "hands-on" experience with the manipulation of visual and other sensory images. Thus the process does not have a clean "algorithmic or nonalgorithmic" status.

For this account of "insight" to be acceptable as a viable possibility it must be placed in a context where the close connection between experiential processes and brain processes is understood in naturalistic terms. Otherwise, there will be an impulse to resolve this gap in understanding by reverting, as Penrose did, to prequantum ways of thinking. That is, we need the very sort of theory of consciousness that Penrose said he had in mind, and that I endeavor to deliver.

1.12 Comparison with Lockwood

The second of the two books is Michael Lockwood's *Mind, Brain, and the Quantum*. Lockwood argues strongly that traditional philosophical approaches to the problem of the connection between mind and matter suffer seriously from inflexible adherence to a "material" conceptualization of matter that is not supported by scientific knowledge:

> This prejudice in favour of the material seems to me devoid of any sound scientific foundation. Quantum mechanics has robbed matter of its conceptual as much as its literal solidity. Mind and matter are alike in being profoundly mysterious, philosophically speaking. And what the mind–body problem calls for, almost certainly, is a *mutual* accommodation: one which involves conceptual adjustment on both sides of the mind–body divide.[43]

Lockwood argues that just as philosophers concerned with the mind–body problem are, in large measure, inclined to leave the concept of matter in an antiquated state, so physicists concerned with the central question of "measurement" are, in large measure, unwilling to deal with the problem of mind:

> What the quantum-mechanical measurement problem is really alerting us to, I shall argue, is a deep problem as to how *consciousness* (specifically the consciousness of the observer) fits into, or maps on to, the physical world. And that, of course, is the question that lies at the heart of the traditional philosophical mind–body problem.[44]

Lockwood and I are in close agreement on these basic points. However, we adopt technically different stances as regards the status of the random process **R** mentioned above. In order to avoid the puzzling instantaneous influences, and to maintain, at the basic level, various symmetry properties that quantum theory exhibits, he assumes that **R** is not a physically real process, but is rather a sort of illusion, arising from the fact that "consciousness" is continually separating into noncommunicating branches: each "I" separates, from time to time, into a collection of "I s" each with a different ongoing succession of experiences. This is the so-called "many-worlds" or "many-minds" interpretation of quantum theory, and it has many things to recommend it. It has also many technical problems to overcome. Much of Lockwood's book is an attempt to describe, and to some extent deal with, these problems. This shifts the focus of his book generally away from the sort of questions that I address. But Lockwood's book, like Penrose's, gives a glimpse into the philosophical and historical setting that underlies the works assembled here.

1.13 Comparison with Eccles

John Eccles has proposed a theory of the relationship between mind and matter that makes an appeal to quantum theory.[44] That appeal is minimal: it exploits the breaking of determinism that quantum theory entails, without going into the tremendous changes in the conception of matter that constitutes the core of the theory.

The basic aim of Eccles's theory is quite different from that of the present works, which is to see how far one can go in explaining the data pertaining to the mind–matter relationship *without* bringing in any structural forms that are not represented within the quantum-mechanical description of the brain. Eccles's theory, on the contrary, brings in a kind of "soul", which constitutes a "knower" and "controller" of what is going on in the brain. The thoughts reside in this nonphysical structure, rather than being an integral part of the brain itself. The question of how the "knower" is able to transcribe the complex patterns of neural firings into thoughts is not really tackled in any serious way: the "knower" would seem to have to possess a tremendous analytical capacity comparable to that of the brain itself. This would involve an uneconomical redundancy in nature: two computers, one material, and one in some mysterious other realm. The aim of the present works is precisely to show how in a quantum-mechanical nature, which therefore contains among its characteristics both mindlike and matterlike qualitites, one single computer, the mind/brain, suffices.

References

1 A. Einstein, B. Podolsky, and N. Rosen, Is Quantum Mechanical Description of Physical Reality Complete?, *Phys. Rev.* **47**, 777–780 (1935).
2 L. Hardy, *Phys. Rev. Lett.* **71**, 1665– (1993).
3 W. James, *The Principles of Psychology* (Dover, New York, 1950; reprint of 1890 text), vol. 1, p. 1.
4 Ref. 3, p. 3.
5 Ref. 3, p. 3.
6 Ref. 3, p. 4.
7 Ref. 3, p. 129.
8 Ref. 3, p. 129.
9 Ref. 3, p. 130.

10 Ref. 3, p. 134.
11 Ref. 3, pp. 134–135.
12 Ref. 3, p. 135.
13 Ref. 3, p. 135.
14 Ref. 3, pp. 135–136.
15 Ref. 3, p. 136.
16 Ref. 3, p. 276.
17 Ref. 3, pp. 176–177.
18 J. B. Watson, Psychology as the Behaviorist Views It, *Psychol. Rev.* **20**, 158–177 (1913).
19 Ref. 18.
20 A. Einstein, in *Albert Einstein: Philosopher-Scientist*, edited by P. A. Schilpp (Tudor, New York, 1951), p. 87.
21 W. Heisenberg, *Tradition in Science* (Seabury Press, New York, 1983), p. 114.
22 D. Bohm, A Suggested Interpretation of the Quantum Theory in Terms of "Hidden" Variables, *Phys. Rev.* **85**, 166–193 (1952).
23 D. C. Dennett, *Consciousness Explained* (Little, Brown and Company, Boston, 1991), p. 33.
24 B. Baars, *A Cognitive Theory of Consciousness* (Cambridge University Press, Cambridge, 1988), p. 8.
25 W. James, *The Principles of Psychology* (Dover, New York, 1950; reprint of 1890 text), vol. 1, p. 400.
26 G. Ryle, *The Concept of Mind* (Barnes and Noble, New York, 1949).
27 Ref. 23, p. 33.
28 Ref. 23, p. 35.
29 Ref. 23, p. 107.
30 Ref. 23, p. 253.
31 Ref. 23, p. 363.
32 Ref. 23, p. 363.
33 Ref. 23, p. 363.
34 Ref. 23, p. 366.
35 Ref. 23, p. 257.
36 R. Penrose, *The Emperor's New Mind* (Oxford University Press, New York, 1989), p. 438.
37 Ref. 36, p. 423.
38 Ref. 36, p. 423.
39 Ref. 36, p. 281.
40 Ref. 36, p. 10.
41 Ref. 36, p. 10.
42 Ref. 36, p. 423.
43 M. Lockwood, *Mind, Brain, and the Quantum* (Basil Blackwell, Cambridge MA, 1989), p. x.
44 Ref. 43, p. x.
45 J. Eccles, *How the Mind Controls the Brain* (Springer, Berlin Heidelberg, 1993).

2 A Quantum Theory of Consciousness

2.1 Introduction

Classical physics has no natural place for consciousness. According to the classical precepts, the sole ingredients of the physical universe are particles and local fields, and every physical system is completely described by specifying the dispositions in space and time of these two kinds of localizable parts. Furthermore, the dispositions of these parts at early times determine, through certain "laws of motion", their dispositions at all times. The system is logically complete in the sense that it does not logically require, for its description of nature, any things beyond the dispositions of the particles and local fields.

The two cited features of classical physics, namely its local-reductionistic and deterministic aspects, do not entail that there can be no conglomerates that act cohesively as unified wholes. Nor do they entail that such conglomerates cannot control in large measure the motions of their own parts. But these two features of classical physics do entail that, to the extent that classical physics is valid, the motions of material things can be controlled only by things that are themselves deterministically controlled, and, moreover, dynamically equivalent to the forces of classical physics. In particular, because subjective conscious experience is not logically entailed by the concepts of classical physics, any control over brain activity exercised by a conscious experience is, to the extent that classical physics is valid, dynamically equivalent to the control exercised by the classical forces. This equivalence renders conscious experience superfluous, in the sense that the evolution of the physical universe would be exactly the same whether subjective conscious experience exists or not.

The condition "to the extent that classical physics is valid" is critical. It is not satisfied in nature. Classical physics is unable to explain the basic properties of materials, even in inorganic, nonliving, unconscious systems. Yet the operation of the brain depends critically upon the subtle properties of the tissues that make it up. Hence there is no scientific basis for supposing that classical physical theory could provide an adequate conceptual foundation for understanding the dynamics of the mind–brain system. On the other hand, there are ample philosophical reasons to reject the notion that classical physical theory is adequate for this task. Without going here into these reasons I merely cite the complete failure of the three-century-old effort to reconcile the properties of mind with the concepts of classical physics.

Scientists other than quantum physicists often fail to comprehend the enormity of the conceptual change wrought by quantum theory in our basic conception of the nature of matter. For example, it has been claimed, in connection with the mind–brain problem, that the switch to the quantum ideas is "incremental". That is hardly the case. The shift is from a local, reductionistic, deterministic conception of nature in which consciousness has no logical place, and can do nothing but passively watch a preprogrammed course of events, to a nonlocal, nonreductionistic, nondeterministic, conception of nature in which there is a perfectly natural place for consciousness, a place that allows each conscious event, conditioned, but not bound, by any known law of nature, to grasp a possible large-scale metastable pattern of neuronal activity in the brain, and convert its status from "possible" to "actual".

Two revisions in physics lead to the possibility of this profound change in the role of subjective conscious experience in mind–brain dynamics. The first is the opening up, by Heisenberg's indeterminacy principle, of at least the logical possibility that some entity not strictly controlled by the mechanical laws of physics *could* exercise supervenient downward control over the course of physical events. The second is the introduction into physics of physical events that are appropriate counterparts to conscious events, in the critical sense that each such physical event can actualize, as a whole, a complex large-scale metastable pattern of physical activity generated within a complex physical system by the action of the mechanical laws.

2.2 Heisenberg's Picture of the Physical World

According to the strictly orthodox view, quantum theory provides no ordinary sort of picture of the physical world itself. Its principal founders, Bohr and Heisenberg, insisted that the theory must, strictly speaking, be viewed as merely a set of rules for making predictions about observations obtained under certain special kinds of experimental conditions.[1] The detailed form of these quantum predictions is such as to render quantum theory logically incompatible with any local-reductionistic physical world of the kind postulated in classical physics.[2] However, Heisenberg did eventually offer a highly nonclassical kind of picture of the physical world itself.[3] Heisenberg's picture may not be the only possible conception of nature compatible with the predictions of quantum theory,[4] but it is certainly a possible one, and it is, I believe, the image currently favored by the majority of the practicing quantum physicists who allow themselves the luxury of a coherent conception of the physical world itself.

My proposal regarding consciousness is based on Heisenberg's picture of the world, or, more accurately, upon my elaboration upon his picture, which he did not describe in great detail. The central idea in Heisenberg's picture of nature is that atoms are not "actual" things. The physical state of an atom, or of an assembly of atoms, represents only a set of "objective tendencies" for certain peculiar kinds of "actual events" to occur. These events are things of a new and entirely different kind. Moreover, the fundamental dynamical process of nature is no longer one single uniform process, as it is in classical physics. It consists rather of two different processes. One of these processes is a continuous, orderly, deterministic evolution. This process is controlled by fixed mathematical laws that are direct generalizations of the laws of classical physics. However, this process does not control the actual things themselves. It controls only the propensities, or objective tendencies, for the occurrence of the actual things. The other dynamical process consists of a sequence of unruly "quantum jumps". These jumps are not individually controlled by any known law of physics. Yet collectively they conform to strict statistical rules. These quantum jumps are considered to be the "actual" things in nature. They are Heisenberg's actual events.

Heisenberg described his picture of the world in connection with the behavior of a quantum measuring device. In that context it is important to recognize that quantum theory naturally accommodates transformations of variables. Thus in the description of large objects one need not use directly the coordinates of the individual particles. It is often more useful to introduce variables that represent various "observable" features of the object.

Our direct sensory perceptions of a macroscopic object containing a huge number of particles can be represented by a relatively small number of "observable" variables. Each of these variables can be confined by the data obtained by our direct sensing of the object only within an interval that is generally so large that quantum effects become irrelevant. Of course, one might try to use some device to probe those features not describable in terms of these observable variables, but then our direct sensory impressions would be of the observable characteristics of that device. Thus we human beings are effectively imprisoned in the physical world described by observable variables: we can access the rest of the physical world only through this extremely limited set of variables. This fact is crucial to the application of quantum theory.

In the typical measurement situation discussed by Heisenberg there is a measuring device that is being used to measure some property of an atomic-sized quantum system. The device must be in a state of unstable equilibrium, so that a small signal from the atomic-sized system can trigger a chain of events leading to a change of certain observable features of the device.

In this situation there is the possibility of a change of the observable macroscopic state of the device from one metastable configuration to another. Here Heisenberg introduces his key idea, the notion of an "actual event". The possibility of introducing into physical theory this new concept of an actual event arises from the fact that the deterministic part of the quantum dynamics is expressed in terms of a quantity that, from a mathematical point of view, ought to represent probabilities. Yet within the mathematics itself there is no clear indication of exactly what these probabilities refer to—what these probabilities are probabilities *of*.

Heisenberg supplied an answer by proposing, in effect, that certain probabilities defined by the theory be interpreted as the "objective tendencies", or propensities, for corresponding *actual events* to occur. Each of these actual events is the actualization of one of the distinct metastable configurations of the observable degrees of freedom generated by the mechanical laws of motion, and the eradication of all those remaining patterns of physical activity that might have been actualized, but were not.

The introduction of these actual events carries quantum theory far beyond the ontologically neutral stance of the strictly orthodox interpretation. In the orthodox interpretation the quantum probabilities are interpreted as simply the probabilities that the community of human observers will "observe" particular ones of these distinct metastable states. The difference between this orthodox interpretation in terms of observations and Heisenberg's ontological interpretation in terms of actual events is, at the practical level, completely negligible in all experimental situations that have yet been examined. Yet there is an important theoretical difference: Heisenberg's picture allows quantum theory to be viewed as a coherent description of the evolution of physical reality itself, rather than merely a set of stark statistical rules about connections between human observations.

2.3 Brain Dynamics

The human brain is a device that can process sensory inputs, formulate possible responses to the sensed situation, select a response, and oversee the execution of that response. This activity is dependent on the momentary physical state of the brain, which is a product of many factors, such as genetic structuring, conditioning, learning, and self-organization (e.g., reflection), among many others. The brain contains a huge network of neurons linked at synapses. These synaptic links allow electrical pulses in neurons to tend to produce or inhibit similar pulses in other neurons. The complex feed-back and feed-forward linkages allow the occurrence of an immense number of alternative possible metastable reverberating patterns of neural pulses. The persistence for a short time of such a pattern apparently conditions the synaptic junctions in a way that facilitates the excitation of this pattern as a component of subsequent metastable patterns of reverberation.[5]

In the formulation and execution of a bodily response a key role must be played by the *body schema*, which is the brain's representation of the dispositions of the parts of the body that it is supervising. This body schema is associated with an *external-world schema*, which is the brain's representation of the environment of the body that is represented by the body schema. These two schema are essentially stable: they do not change spontaneously; they are changed only by a particular process, which replaces the "current" schema by a new one, and places the old one into an appropriate slot in a *historical schema*.

In addition to the body schema and the external-world schema there is a *belief schema*, and these three representations are parts of the *"self and world" schema*. This latter schema lies at the "current" end of a general historical schema, into which each "self and world" schema is placed when it is displaced by a new one.

2.4 Consciousness

My proposal for identifying conscious events with certain specific kinds of brain events in Heisenberg's quantum-mechanical picture of physical reality is based upon three observations:

1   The schema described above are "classical" in the sense that they can be examined and manipulated in ways analogous to the ways that we examine and manipulate macroscopic objects: the schema are not appreciably

disturbed by a mere examination, and they can be "manipulated" by appropriate kinds of processing. It is therefore reasonable to suppose that these schema are represented by physical structures that are describable in terms of variables of the type that in measuring devices were called *observables*. In fact, these brain structures are the *only* structures that can ever really be "observed": sensory inputs must be converted into the external-world schema (including affiliated buffers) before they can be perceived.

2 Brain processes involve chemical processes, and hence must, in principle, be treated quantum mechanically. In particular, the transmission process occurring at a synaptic junction is apparently triggered by the capture of a small number of calcium ions at an appropriate release site. In a quantum-mechanical treatment, the locations of these calcium ions must be represented by a probability function. This effectively smears these particles over large regions, in a quantum-statistical sense. Thus the question of whether or not a given synapse will transmit a signal is a problem that must be treated quantum mechanically: a quantum-mechanical component must be added to the other uncertainties, such as those generated by thermal noise, that enter into the decision as to whether or not the synapse will fire.

There are hundreds of billions of synapses coupled together in a highly nonlinear fashion. And there must be a huge number of metastable reverberating patterns of pulses into which the brain might evolve.

Computer simulations of brain networks in the classical case indicate that the final stable state into which a brain evolves is strongly dependent upon the synaptic parameters.[6] Although analogous computations are needed for the quantum case it appears to me exceedingly probable, by virtue of (1) the inherent sensitivity of nonlinear systems of this kind to variations in parameters, (2) the strong dependence of the process at the synaptic junction upon the locations of small numbers of calcium ions, and (3) the large number of possible metastable states into which the brain might evolve, that, in the absence of any quantum jumps, a brain will generally evolve quantum mechanically from one metastable configuration into a *quantum superposition* of many metastable configurations, and sometimes into a superposition that ascribes non-negligible quantum probabilities to several alternative possible metastable states of the "self and world" schema. Note that the fatigue characteristics of the synaptic junctions will cause any given metastable pattern to become, after a short time, unstable:[5] the system will thus be forced to search for a new metastable configuration, and will therefore continue to evolve, if unchecked by a quantum jump, into a superposition of states characterized by increasingly disparate self and world schemas.

3   The situation described above is, from the physical point of view, essentially the same as the one considered by Heisenberg, with the human brain in place of his measuring device. Thus if one accepts his picture of the world, then one must accept also that if the brain evolves into a superposition of states characterized by different "self and world" schema then an actual event must select and actualize one of these "observable" states, and eradicate the others. I propose to identify each such actual brain event with a conscious event, and, conversely, to identify each conscious event with an actual brain event of this particular kind.

The only relativistically invariant way to represent a Heisenberg actual event is by a change in the Heisenberg state of the universe. In the interim between actual events there is, in the Heisenberg picture, only a global structure of potentialities that extends uniformly over all of spacetime. There is, in keeping with the special theory of relativity, no structure that connects a spacetime point to another point that is "simultaneous with it" in any favored physical sense. However, each actual event is localized: each actual event is associated with a local spacetime region in which a certain classically describable metastable pattern of activity is actualized. This event is represented by a sudden jump in the global Heisenberg state of the universe.

Within the general framework of the Heisenberg picture an actual event *could* occur already at the level of the firing of an individual neuron: an actual event *could* fix whether a certain individual neuron does or does not fire. However, von Neumann's analysis of the process of measurement shows that the actual events in the brain *need not* occur at the level of the individual neurons: an actual event can perfectly well actualize *the entire large-scale integrated pattern of neural excitations associated with the metastable state of the brain that goes along with a particular conscious thought*. Indeed, von Neumann's words seem to suggest that *all* actual events are events of this kind.[7] However, in the Heisenberg ontology adopted here the actual events are not *exclusively* conscious events. On the other hand, every conscious event is an actual event: it is an event that *selects one of the alternative possible high-level metastable configurations of brain activity from among the host of such patterns mechanically generated by the Schrödinger equation*. Each conscious event corresponds, therefore, to an entity that *supervenes* over the quantum-mechanical laws analogous to the laws of classical physics: the conscious event corresponds to a Heisenberg event that actualizes the classically describable metastable quantum state of the brain that represents this conscious experience in the physicist's description of nature.

2.5 Remarks

1   The purpose of conscious thought is to guide the organism. This must be done by forming a projection into the future, and, more specifically, by forming a projected "self and world" schema. Thus, one step ahead of the current "self and world" schema is the *projected* "self and world" schema. This is the thing that is selected by a conscious act. It is the template that directs the subconscious processes that control, among other things, the motor activities.

2   The physical event is "functionally equivalent" to the corresponding psychological event. The physical event selects a projected "self and world" schema that acts as the template for brain action, whereas the corresponding psychological event selects the associated imagined projected "self and world". Thus the identification of these events is neither ad hoc nor arbitrary: it is an expression of their functional equivalence.

3   To justify this claim of "equivalence" an isomorphism must be established between the intrinsic structure of a conscious thought, as it is described by psychologists such as James, and the intrinsic structure of the "projected self and world schema", which is the template that directs the unconscious processes of the brain in the way specified by that conscious thought. This key issue is addressed in reference 8.

4   The model shows how experiences exhibiting the empirically established features of conscious experience can arise essentially automatically out of quantum theory, provided the brain operates in the way suggested by Heisenberg's picture of nature. The theory is predictive in that it entails that brain process must be controlled by a top-level process having the specific dynamical and structural features, expressed in terms of self and world schema and memory, described in reference 8.

References

1   H. P. Stapp, The Copenhagen Interpretation, *Am. J. Phys.* **40**, 1098 (1972) and chap. 3 of the present book.

2   H. P. Stapp, Quantum Nonlocality and the Description of Nature, in *Philosophical Consequences of Quantum Theory*, edited by J. Cushing and E. McMullin (Notre Dame University Press, 1989), pp. 154–174; EPR and Bell's Theorem: A Critical Review, *Found. Phys.* **21**, 1 (1991); Noise-Induced Reduction of Wave Packets and Faster-than-Light Influence, *Phys. Rev. A* **46**, 6860 (1992); (with D. Bedford), Bell's Theorem in an Indeterministic Universe, submitted to *Synthese*.

3   W. Heisenberg, *Physics and Philosophy* (Harper and Row, New York, 1958), chap. III.

4 H. P. Stapp, Quantum Theory and Emergence of Patterns in the Universe, in *Bell's Theorem, Quantum Theory, and Conceptions of the Universe*, edited by Menas Kafatos (Kluwer, Dordrecht, Boston, 1989).

5 J. I. Hubbard, Mechanism of Transmitter Release, *Prog. Biophys. Mol. Biol.* **21**, 33 (1970).

6 L. Ingber, Statistical Mechanics of Neocortical Interactions. Dynamics of Synaptic Modification, *Phys. Rev. A* **28**, 395–416 (1983); Statistical Mechanics of Neocortical Interactions. Derivation of Short-Term-Memory Capacity, *Phys. Rev. A* **29**, 3346–3358 (1984).

7 J. von Neumann, *Mathematical Foundations of Quantum Mechanics* (Princeton University Press, Princeton, 1955); E. Wigner, Remarks on the Mind–Body Problem, in *The Scientist Speculates*, edited by I. J. Good (Heinemann, London, and Basic Books, New York, 1962).

8 H. P. Stapp, A Quantum Theory of the Mind–Brain Interface, Lawrence Berkeley Laboratory Report LBL-28574 Expanded, University of California, Berkeley, 1991, and chap. 5 of the present book.

Part II

Theory

3 The Copenhagen Interpretation

3.1 Introduction

Scientists of the late 1920s, led by Bohr and Heisenberg, proposed a conception of nature radically different from that of their predecessors. The new conception, which grew out of efforts to comprehend the apparently irrational behavior of nature in the realm of quantum effects, was not simply a new catalog of the elementary spacetime realities and their modes of operation. It was essentially a rejection of the presumption that nature could be understood in terms of elementary spacetime realities. According to the new view, the complete description of nature at the atomic level was given by probability functions that referred not to underlying microscopic spacetime realities but rather to the macroscopic objects of sense experience. The theoretical structure did not extend down and anchor itself on fundamental microscopic spacetime realities. Instead it turned back and anchored itself in the concrete sense realities that form the basis of social life.

This radical concept, called the Copenhagen interpretation, was bitterly challenged at first but became during the 1930s the orthodox interpretation of quantum theory, nominally accepted by almost all textbooks and practical workers in the field.

Recently, perhaps partly in response to the severe technical difficulties now besetting quantum theory at the fundamental level, there has been mounting criticism of the Copenhagen interpretation. The charges range from the claim that it is a great illogical muddle to the claim that it is in any case unnecessary, and hence, in view of its radical nature, should be rejected. Reference 1 contains some stoutly worded attacks on the Copenhagen interpretation. Reference 2 is a more moderately worded review article that firmly rejects the Copenhagen interpretation. Reference 3 is a list of articles in the physical literature that espouse a variety of views on the question.

The striking thing about these articles is the diversity they reveal in prevailing conceptions of the Copenhagen interpretation itself. For example, the picture of the Copenhagen interpretation painted in reference 1 is quite different from the pictures painted in references 2 and 3 by practicing physicists. And these latter pictures themselves are far from uniform.

The cause of these divergences is not hard to find. Textbook accounts of the Copenhagen interpretation generally gloss over the subtle points. For clarification readers are directed to the writings of Bohr[4] and Heisenberg[5]. Yet clarification is difficult to find there. The writings of Bohr are extraordinarily elusive. They rarely seem to say what you want to know. They weave a web of words around the Copenhagen interpretation but do not say exactly what it is. Heisenberg's writings are more direct. But his way of speaking suggests a subjective interpretation that appears quite contrary to the apparent intentions of Bohr. The situation is perhaps well summarized by von Weizsäcker, who, after expressing the opinion that the Copenhagen interpretation is correct and indispensable, says he must

> add that the interpretation, in my view, has never been fully clarified. It needs an interpretation, and that will be its only defense.[6]

Von Weizsäcker is surely correct. The writings of Bohr and Heisenberg have, as a matter of historical fact, not produced a clear and unambiguous picture of the basic logical structure of their position. They have left impressions that vary significantly from reader to reader. For this reason a clarification of the Copenhagen interpretation is certainly needed. My aim here is to provide one. More precisely, my aim is to give a clear account of the logical essence of the Copenhagen interpretation. This logical essence should be distinguished from the inhomogeneous body of opinions and views that now constitute the Copenhagen interpretation itself. The logical essence constitutes, I believe, a completely rational and coherent position.

The plan of the work is as follows. First, quantum theory is described from the point of view of actual practice. Then, to provide contrast, several non-Copenhagen interpretations are considered. Next, to provide background, some philosophical ideas of William James are introduced. The pragmatic character of the Copenhagen interpretation is then discussed, and the incompatibility of the completeness of quantum theory with the external existence of the spacetime continuum of classical physics is noted. Finally, the question of the completeness of quantum theory is examined.

3.2 A Practical Account of Quantum Theory

Quantum theory is a procedure by which scientists predict probabilities that measurements of specified kinds will yield results of specified kinds in situations of specified kinds. It is applied in circumstances that are described by saying that a certain physical system is first prepared in a specified manner and is later examined in a specified manner. And this examination, called a measurement, is moreover such that it can yield, or not yield, various possible specified results.

The procedure is this: The specifications A on the manner of preparation of the physical system are first transcribed into a wave function $\Psi_A(x)$. The variables x are a set of variables that are characteristic of the physical system being prepared. They are called the degrees of freedom of the prepared system. The description of the specifications A is couched in a language that is meaningful to an engineer or laboratory technician. The way in which these operational specifications A are translated into a corresponding wave function $\Psi_A(x)$ is discussed later.

The specifications B on the subsequent measurement and its possible result are similarly couched in a language that allows a suitably trained technician to set up a measurement of the specified kind and to determine whether the result that occurs is a result of the specified kind. These specifications B on the measurement and its result are transcribed into a wave function $\Psi_B(x)$, where y is a set of variables that are called the degrees of freedom of the measured system.

Next a transformation function $U(x; y)$ is constructed in accordance with certain theoretical rules. This function depends on the type of system that was prepared and on the type of system that was measured, but not on the particular wave functions $\Psi_A(x)$ and $\Psi_B(y)$. The "transition amplitude"

$$\langle A|B \rangle = \int \Psi_A(x)U(x; y)\Psi_B^*(y)\mathrm{d}x\mathrm{d}y$$

is computed. The predicted probability that a measurement performed in the manner specified by B will yield a result specified by B, if the preparation is performed in the manner specified by A, is given by

$$P(A, B) = |\langle A|B \rangle|^2.$$

The experimental physicist will, I hope, recognize in this account a description of how he uses quantum theory. First he transforms his information about the preparation of the system into an initial wave function. Then he applies to it some linear transformation, calculated perhaps from the Schrödinger equation, or perhaps from the S matrix, which converts the initial wave function into a final wave function. This final wave function, which is built on the degrees of freedom of the measured system, is then folded into the wave function corresponding to a possible result. This gives the transition amplitude, which is multiplied by its complex conjugate to give the predicted transition probability.

In a more sophisticated calculation one might use density matrices $\rho_A(x'; x'')$ and $\rho_B(y'; y'')$ instead of $\Psi_A(x)$ and $\Psi_B(y)$ to represent the prepared system and the possible result. This would allow for preparations and measurements that correspond to statistical mixtures. But this generalization could be obtained also by simply performing classical averages over various $\Psi_A(x)$ and $\Psi_B(y)$.

The above account describes how quantum theory is used in practice. The essential points are that attention is focused on some system that is first prepared in a specified manner and later examined in a specified manner. Quantum theory is a procedure for calculating the predicted probability that the specified type of examination will yield some specified result. This predicted probability is the predicted limit of the relative frequency of occurrence of the specified result, as the number of systems prepared and examined in accordance with the specifications goes to infinity.

The wave functions used in these calculations are functions of a set of variables characteristic of the prepared and measured systems. These systems are often microscopic and not directly observable. No wave functions of the preparing and measuring devices enter into the calculation. These devices are described operationally. They are described in terms of things that can be recognized and/or acted upon by technicians. These descriptions refer to the macroscopic properties of the preparing and measuring devices.

The crucial question is: How does one determine the transformations $A \rightarrow \Psi_A$ and $B \rightarrow \Psi_B$? These transformations transcribe procedural descriptions of the manner in which technicians prepare macroscopic objects, and recognize macroscopic responses, into mathematical functions built on the degrees of freedom of the (microscopic) prepared and measured systems. The problem of constructing this mapping is the famous "problem of measurement" in quantum theory.

The problem of measurement was studied by von Neumann.[7] He begins with the idea that one should describe the combined system composed of the original systems plus the original measuring devices in terms of a quantum-mechanical wave function, and use quantum theory itself to calculate the needed mappings. This program has never been carried out in any practical case. One difficulty is that actual macroscopic devices are so complicated that qualitative calculations lie beyond present capabilities. The second problem is that such calculations would, in any case, provide only connections between the wave functions Φ of the preparing and measuring devices and the wave functions Ψ of the original system. There would remain the problem of finding the mappings $A \rightarrow \Phi_A$ and $B \rightarrow \Phi_B$.

Von Neumann's approach is not the one that is adopted in actual practice; no one has yet made a qualitatively accurate theoretical description of a measuring device. Thus what experimentalists do, in practice, is to calibrate their devices.

Notice, in this connection, that if one takes N_A different choices of A and N_B different choices of B, then one has only $N_A + N_B$ unknown functions A and B, but $N_A \times N_B$ experimentally determinable quantities $|\langle A|B \rangle|^2$. Using this leverage, together with plausible assumptions about smoothness, it is possible to build up a catalog of correspondences between what experimental physicists do and see, and the wave functions of the prepared and measured systems. It is this body of accumulated empirical knowledge that bridges the gap between the operational specifications A and B and their mathematical images Ψ_A and Ψ_B.

The above description of how quantum theory is used in practice will be used in the account of the Copenhagen interpretation. Before describing that interpretation itself I shall, to provide contrast, describe several other approaches.

3.3 Several Other Approaches

3.3.1 The Absolute-Ψ Approach

Von Neumann's lucid analysis of the process of measurement is the origin of much of the current worry about the interpretation of quantum theory. The basic worrisome point can be illustrated by a simple example.

Suppose a particle has just passed through one of two slits. And suppose a 100%-efficient counter is placed behind each slit, so that, by seeing which counter fires, a human observer can determine through which slit the particle passed.

Suppose the particle is represented initially by a wave function that assigns equal probabilities to the parts associated with the two slits. And consider a quantum-theoretical analysis of the process of measurement in which both the particle and the two counters are represented by wave functions.

It follows directly and immediately from the superposition principle (i.e., linearity) that the wave function of the complete system after the measurement necessarily will consist of a superposition of two terms. The first term will represent the situation in which (1) the particle has passed through the first counter, (2) the first counter has fired, and (3) the second counter has not fired. The second term will represent the situation in which (1) the particle has passed through the second counter, (2) the second counter has fired, and (3) the first counter has not fired. These two terms evolve from the two terms in the wave function of the initial particle. The presence of both terms is a direct and unavoidable consequence of the superposition principle, which ensures that the sum of any two solutions of the equation of motion is another solution.

Notice now that the counters are macroscopic objects and that the wave function necessarily contains a sum of two terms, one of which corresponds to the first counter's having fired but not the second, and the other of which corresponds to the second counter's having fired but not the first. Thus the wave function necessarily corresponds to a sum of two logically incompatible *macroscopic* possibilities.

To dramatize this situation, suppose the human observer now looks at the counters and runs upstairs or downstairs depending on which counter he sees firing. Then the wave function of the entire system of particle plus counters plus human observer will consist, eventually, of a sum of two terms. One term will represent the human observer running upstairs, and the other term will represent this same human observer running downstairs. Both terms must necessarily be present in the wave function, simply by virtue of the superposition principle.

This fact that the wave function necessarily develops into a sum of parts that correspond to incompatible *macroscopic* possibilities must be squared with the empirical facts. The human observer does not run both upstairs and downstairs. He does one or the other, not both. Therefore the wave function must collapse to a form that is consistent with what actually does happen. But such a collapse is definitely incompatible with the superposition principle.

This violation of the superposition principle bothers some thinkers. Wigner calls the existence of the two modes of change of the wave function — i.e., the smooth causal evolution and the fitful statistical jumps associated with measurements — a strange dualism, and says that the probabilistic behavior is almost diametrically opposite to what one would expect from ordinary experience.[8] He and Ludwig speculate that quantum theory may have to be modified by the addition of a nonlinear effect in the macroscopic realm in order to arrive at a consistent theory of measurements.[9] Wigner even speculates that the nonlinearity may be associated with the action of mind on matter.[10]

An even more radical proposal was made by Everett[11] and supported by Wheeler[12] and Bryce DeWitt[13]. According to this proposal the human observer actually runs both upstairs and downstairs at the same time. When the human observer sees the counter fire he breaks into two separate editions of himself, one of which runs upstairs while the other runs down. However, the parts of the wave function corresponding to these two different possibilities move into different regions of the multiparticle configuration space and consequently do not interfere. Therefore the two editions will never be aware of each other's existence. Thus appearances are saved without violating the superposition principle.

This proposal is, I think, unreasonable. A wave function times its complex conjugate has the mathematical properties of a probability function. Probability functions for composite systems are naturally defined on the product of the spaces of the individual component systems; it is this property that allows different statistical weights to be assigned to the various logically alternative possibilities. A decomposition of a wave function into parts corresponding to different logical alternatives is thus completely natural. In the example described—with the initial specification as described there—there is a finite probability that the observer will be running upstairs, and a finite probability that he will be running downstairs. Thus the wave function necessarily must have both parts. If it collapsed to one part or the other, it would no longer correctly describe the probabilities corresponding to the original specifications.

Of course, if the original specifications are replaced by new ones that include now the specification that the observer is running upstairs, not downstairs, then the original wave function will naturally be replaced by a new one, just as it would be in classical statistical theory.

In short, the mathematical properties of the wave functions are completely in accord with the idea that they describe the evolution of the *probabilities* of the actual things, not the actual things themselves. The idea that they describe also the evolution of the actual things themselves leads to metaphysical monstrosities. These might perhaps be accepted if they were the necessary consequences of irrefutable logic. But this is hardly the case here. The basis of Everett's whole proposal is the premise that the superposition principle cannot suddenly fail. This premise is sound. But the natural and reasonable conclusion to draw from it is that the wave functions describe the evolution of the *probabilities* of the actual things, not the evolution of the actual things themselves. For the mathematical form and properties of the wave function, including its lawful development in accordance with the superposition principle, are completely in accord with the presumption that it is a probability function. The addition of the metaphysical assumption that the wave function represents the evolution of not only the probabilities of the actual things, but of also the actual things themselves, is unreasonable because its only virtue is to save the superposition principle, which, however, is not in jeopardy unless one introduces this metaphysical assumption.

Everett's proposal, and also those of Wigner and Ludwig, are the outgrowth of a certain tendency to ascribe to the wave function a quality of absoluteness that goes beyond what is normally and naturally attached to a probability function. This tendency can perhaps be traced to what Rosenfeld calls "a radical difference in conception (going back to von Neumann) . . .",[14] this radical difference being with the ideas of Bohr. Von Neumann's application of quantum theory to the process of measurement itself, coupled with his parallel treatments of the two very different modes of development of the wave function—i.e., the smooth dynamical evolution, and the abrupt changes associated with measurement—tend to conjure up the image of some absolute wave function developing in time under the influence of two different dynamical mechanisms. The living, breathing scientist who changes the wave function he uses as he receives more information is replaced by a new dynamical mechanism. The resulting picture is strange indeed.

In the Copenhagen interpretation the notion of an absolute wave function representing the world itself is unequivocally rejected. Wave functions, like the corresponding probability functions in classical physics, are associated with the studies by scientists of finite systems. The devices that prepare and later examine such systems are regarded as parts of the ordinary classical physical world. Their spacetime dispositions are interpreted by the scientist as information about the prepared and examined systems. Only these latter systems are represented by wave functions. The probabilities involved are the probabilities of specified responses of the measuring devices under specified conditions.

New information available to the scientist can be used in two different ways. It can be considered to be information about the response of a measuring device to the system being examined. In this case the probability of this response is the object of interest. On the other hand, the new information can also be regarded as part of the specification of a new preparation. The wave function that represents this new specification will naturally be different from the wave function that represented the original specifications. One would not expect the superposition principle to be maintained in the change of the wave function associated with a change of specifications.

This pragmatic description is to be contrasted with descriptions that attempt to peer "behind the scenes" and tell us what is "really happening". Such superimposed images can be termed metaphysical appendages insofar as they have no testable consequences. The pragmatic interpretation ignores all such metaphysical appendages.

The sharp distinction drawn in this section between probabilities and the actual things to which they refer should not be construed as an acceptance of the real-particle interpretation which is described next.

3.3.2 The Real-Particle Interpretation

The real-particle interpretation affirms that there are real particles, by which is meant tiny localized objects, or disturbances, or singularities, or other things that stay together like particles should, and do not spread out like waves. According to this interpretation, the probability functions of quantum theory describe, typically, the probability that a real particle is in such-and-such a region. This real-particle interpretation is defended by Popper in reference 1, and by Ballentine in reference 2.

Confidence in the existence of real particles was restored by Bohm's illustration of how nonrelativistic Schrödinger theory can be made compatible with the existence of point particles.[15] The price paid for this achievement is this: All the particles in the (model) universe are instantly and forcefully linked together. What happens to any particle in the universe instantly and violently affects every other particle.

In such a situation it is not clear that we should continue to use the term "particle". For the entire collection of "particles" in Bohm's universe acts as a single complex entity. Our usual idea of a particle is an abstraction from experience about macroscopic objects, and it normally carries, as part of the idea of localization, the idea that the localized entity is an independent entity, in the sense that it depends on other things in the universe only through various "dynamical" effects. These dynamical effects are characterized by a certain respect for spacetime separations. In particular, they are "causal". If the connections between particles radically transcend our idea of causal dynamical relationships, then the appropriateness of the word "particle" can be questioned.

Bell has shown that the statistical predictions of quantum theory are definitely incompatible with the existence of an underlying reality (that resembles the observed world at the macroscopic level) whose spatially separated parts are independent realities linked only by causal dynamical relationships.[16] The spatially separated parts of any underlying reality must be linked in ways that completely transcend the realm of causal dynamical connections. The spatially separated parts of any such underlying reality are not independent realities, in the ordinary sense.

Bell's theorem does not absolutely rule out the real-particle interpretation, if one is willing to admit these hyperdynamical connections. But they fortify the opinion that a dynamical theory based on such a real entity would have no testable dynamical consequences. For the strong dependence of individual effects here on Earth upon the fine details of what is happening all over the universe apparently rules out any ordinary kind of test of such a theory.

3.4 The Pragmatic Conception of Truth

To prepare the mind for the Copenhagen interpretation it is useful to recall some ideas of William James.[17] James argued at length for a certain conception of what it means for an idea to be true. This conception was, in brief, that an idea is true if it works.

James's proposal was at first scorned and ridiculed by most philosophers, as might be expected. For most people can plainly see a big difference between whether an idea is true and whether it works. Yet James stoutly defended his idea, claiming that he was misunderstood by his critics.

It is worthwhile to try to see things from James's point of view.

James accepts, as a matter of course, that the truth of an idea means its agreement with reality. The questions are: What is the "reality" with which a true idea agrees? And what is the relationship "agreement with reality" by virtue of which that idea becomes true?

All human ideas lie, by definition, in the realm of experience. Reality, on the other hand, is usually considered to have parts lying outside this realm. The question thus arises: How can an idea lying inside the realm of experience agree with something that lies outside? How does one conceive of a relationship between an idea, on the one hand, and something of such a fundamentally different sort? What is the structural form of that connection between an idea and a transexperiential reality that goes by the name of "agreement"? How can such a relationship be comprehended by thoughts forever confined to the realm of experience?

The contention that underlies James's whole position is, I believe, that a relationship between an idea and something else can be comprehended only if that something else is also an idea. Ideas are eternally confined to the realm of ideas. They can "know" or "agree" only with other ideas. There is no way for a finite mind to comprehend or explain an agreement between an idea and something that lies outside the realm of experience.

So if we want to know what it means for an idea to agree with a reality we must first accept that this reality lies in the realm of experience.

This viewpoint is not in accord with the usual idea of truth. Certain of our ideas are ideas about what lies outside the realm of experience. For example, I may have the idea that the world is made up of tiny objects called particles. According to the usual notion of truth this idea is true or false according to whether or not the world really is made up of such particles. The truth of the idea depends on whether it agrees with something that lies outside the realm of experience.

Now the notion of "agreement" seems to suggest some sort of similarity or congruence of the things that agree. But things that are similar or congruent are generally things of the same kind. Two triangles can be similar or congruent because they are the same kind of thing: the relationships that inhere in one can be mapped in a direct and simple way into the relationships that inhere in the other.

But ideas and external realities are presumably very different kinds of things. Our ideas are intimately associated with certain complex, macroscopic, biological entities—our brains—and the structural forms that can inhere in our ideas would naturally be expected to depend on the structural forms of our brains. External realities, on the other hand, could be structurally very different from human ideas. Hence there is no a priori reason to expect that the relationships that constitute or characterize the essence of external reality can be mapped in any simple or direct fashion into the world of human ideas. Yet if no such mapping exists then the whole idea of "agreement" between ideas and external realities becomes obscure.

The only evidence we have on the question of whether human ideas can be brought into exact correspondence with the essences of the external realities is the success of our ideas in bringing order to our physical experience. Yet the success of ideas in this sphere does not ensure the exact correspondence of our ideas to external reality.

On the other hand, the question of whether ideas "agree" with external essences is of no practical importance. What is important is precisely the success of the ideas—if ideas are successful in bringing order to our experience, then they are useful even if they do not "agree", in some absolute sense, with the external essences. Moreover, if they are successful in bringing order into our experience, then they do "agree" at least with the aspects of our experience that they successfully order. Furthermore, it is only this agreement with aspects of our experience that can ever really be comprehended by man. That which is not an idea is intrinsically incomprehensible, and so are its relationships to other things. This leads to the pragmatic viewpoint that ideas must be judged by their success and utility in the world of ideas and experience, rather than on the basis of some intrinsically incomprehensible "agreement" with nonideas.

The significance of this viewpoint for science is its negation of the idea that the aim of science is to construct a mental or mathematical image of the world itself. According to the pragmatic view, the proper goal of science is to augment and order our experience. A scientific theory should be judged on how well it serves to extend the range of our experience and reduce it to order. It need not provide a mental or mathematical image of the world itself, for the structural form of the world itself may be such that it cannot be placed in simple correspondence with the types of structures that our mental processes can form.

James was accused of subjectivism—of denying the existence of objective reality. In defending himself against this charge, which he termed slanderous, he introduced an interesting ontology consisting of three things: (1) private concepts, (2) sense objects, (3) hypersensible realities. The private concepts are subjective experiences. The sense objects are public sense realities, i.e., sense realities that are independent of the individual. The hypersensible realities are realities that exist independently of all human thinkers.[18]

Of hypersensible realities James can talk only obliquely, since he recognizes both that our knowledge of such things is forever uncertain and that we can moreover never even think of such things without replacing them by mental substitutes that lack the defining characteristics of that which they replace, namely the property of existing independently of all human thinkers.

James's sense objects are curious things. They are sense realities and hence belong to the realm of experience. Yet they are public: they are independent of the individual. They are, in short, objective experiences. The usual idea about experiences is that they are personal or subjective, not public or objective.

This idea of experienced sense objects as public or objective realities runs through James's writings. The experience "tiger" can appear in the mental histories of many different individuals. "That desk" is something that I can grasp and shake, and you also can grasp and shake. About this desk James says:

> But you and I are commutable here; we can exchange places; and as you go bail for my desk, so I can go bail for yours. This notion of a reality independent of either of us, taken from ordinary experience, lies at the base of the pragmatic definition of truth.[19]

These words should, I think, be linked with Bohr's words about classical concepts as the basis of communication between scientists. In both cases the focus is on the concretely experienced sense realities—such as the shaking of the desk—as the foundation of social reality. From this point of view the objective world is not built basically out of such airy abstractions as electrons and protons and "space". It is founded on the concrete sense realities of social experience, such as a block of concrete held in the hand, a sword forged by a blacksmith, a Geiger counter prepared according to specifications by laboratory technicians and placed in a specified position by experimental physicists.

This brief excursion into philosophy provides background for the Copenhagen interpretation, which is fundamentally a shift to a philosophic perspective resembling that of William James.

3.5 The Pragmatic Character
of the Copenhagen Interpretation

The logical essence of the Copenhagen interpretation is summed up in the following two assertions:

1 The quantum-theoretical formalism is to be interpreted *pragmatically*.
2 Quantum theory provides for a *complete* scientific account of atomic phenomena.

Point 1 asserts that quantum theory is fundamentally the procedure described in the practical account of quantum theory given in section 3.2. The central problem for the Copenhagen interpretation is to reconcile this assertion with the claim that it is complete, i.e., to reconcile assertions 1 and 2. This problem is discussed in section 3.7.

The aim of the present section is to document point 1 by the words of Bohr. This fundamental point needs to be definitely settled, for critics often confuse the Copenhagen interpretation, which is basically pragmatic, with the diametrically opposed absolute-Ψ interpretation described in section 3.3. In what follows, particular attention will be paid to the possible conflict of the pragmatic viewpoint with (i) the element of realism in Bohr's attitude toward the macroscopic world, and (ii) any commitment to a fundamental stochastic or statistical element in nature itself.

The quotations from Bohr that follow are taken from his three major works: I. *Atomic Theory and the Description of Nature*; II. *Atomic Physics and Human Knowledge*; and III. *Essays 1958/1962 on Atomic Physics and Human Knowledge*.[4]

The pragmatic orientation of the Copenhagen interpretation is fixed in the opening words of Bohr's first book:

The task of science is both to extend the range of our experience and reduce it to order . . . (I.1)

> In physics . . . our problem consists in the co-ordination of our experience of the external world . . . (I.1)

> In our description of nature the purpose is not to disclose the real essence of phenomena but only to track down as far as possible relations between the multifold aspects of our experience. (I.18)

This commitment to a pragmatic view of science runs through all of Bohr's works. He later links it to the crucial problem of communication:

> As the goal of science is to augment and order our experience, every analysis of the conditions of human knowledge must rest on considerations of the character and scope of our means of communication. (II.88)

> In this connection it is imperative to realize that in every account of physical experience one must describe both experimental conditions and observations by the same means of communication as the one used in classical physics. (II.88)

> The decisive point is to recognize that the description of the experimental arrangement and the recordings of observations must be given in plain language, suitably refined by the usual terminology. This is a simple logical demand, since by the word "experiment" we can only mean a procedure regarding which we are able to communicate to others what we have done and what we have learnt. (III.3)

> . . . we must recognize above all that, even when phenomena transcend the scope of classical physical theories, the account of the experimental arrangement and the recording of observations must be given in plain language, suitably supplemented by technical physical terminology. This is a clear logical demand, since the very word "experiment" refers to a situation where we can tell others what we have done and what we have learned. (II.72)

Bohr's commitment to a pragmatic interpretation of the quantum-mechanical formalism is unambiguous:

> . . . the appropriate physical interpretation of the symbolic quantum-mechanical formalism amounts only to predictions, of determinate or statistical character, pertaining to individual phenomena appearing under conditions defined by classical physical concepts. (II.64)

> . . . the formalism does not allow pictorial representation on accustomed lines, but aims directly at establishing relations between observations obtained under well-defined conditions. (II.71)

> The sole aim of [the quantum-mechanical formalism] is the comprehension of observations obtained under experimental conditions described by simple physical concepts. (II.90)

> Strictly speaking, the mathematical formalism of quantum mechanics and electrodynamics merely offers rules of calculation for the deduction of expectations about observations obtained under well-defined experimental conditions specified by classical physical concepts. (III.60)

Throughout Bohr's writings there is a tacit acceptance of the idea that the external world exists, and that our physical experiences are caused, in part, by the course of external events. This is quite in accord with pragmatism: James admits the existence of hypersensible realities. But there is no commitment by Bohr to the idea that the macroscopic world really is what we naively imagine it to be. The focus is on the descriptions of our physical experiences and the demand that they secure unambiguous communication and objectivity. Referring to the experimental arrangements and observations he says:

> The description of atomic phenomena has in these respects a perfectly objective character, in the sense that no explicit reference is made to any individual observer and that therefore, with proper regard to relativistic exigencies, no ambiguity is involved in the communication of information. As regards all such points, the observation problem of quantum physics in no way differs from the classical physical approach. (III.3)

Bohr's closest approach to a commitment to the idea that the macroscopic world actually is what it appears to be is, I think, the statement:

> The renunciation of pictorial representation involves only the state of atomic objects, while the foundation of the description of the experimental conditions is fully retained. (II.90)

The commitment here is, I believe, to the appropriateness, in quantum theory, of a classical description of the experimental conditions, rather than to the *fundamental* accuracy of classical ideas at the macroscopic level. This position is in complete accord with pragmatism.

In regard to the irreducible statistical element in quantum theory, Bohr was at first ambivalent. An initial acceptance of the notion of a fundamental element of randomness or indeterminism on the part of nature is suggested by the statement:

> . . . we have been forced . . . to reckon with a free choice on the part of nature between various possibilities to which only probability interpretations can be applied . . . (I.4)

However, he soon qualifies this idea (I.19) and later on says that at a Solvay conference

> an interesting discussion arose about how to speak of the appearance of phenomena for which only statistical predictions can be made. The question was whether, as to the occurrence of individual effects, we should adopt the terminology proposed by Dirac, that we were concerned with a choice on the part of "nature", or as suggested by Heisenberg, we should say that we have to do with a choice on the part of the "observer" constructing the measuring instruments and reading their recording. Any such terminology would, however, appear dubious since, on the one hand, it is hardly reasonable

to endow nature with volition in the ordinary sense, while on the other hand it is certainly not possible for the observer to influence the events which may appear under the conditions he has arranged. To my mind there is no other alternative than to admit in this field of experience, we are dealing with individual phenomena and that our possibilities of handling the measuring instruments allow us only to make a choice between the different complementary types of phenomena that we want to study. (II.51)

Later he says:

The circumstance that, in general, one and the same experimental arrangement may yield different recordings is sometimes picturesquely described as a "choice of nature" between such possibilities. Needless to say, such a phrase implies no allusion to a personification of nature, but simply points to the impossibility of ascertaining on accustomed lines directives for the course of a closed indivisible phenomenon. Here, logical approach cannot go beyond the deduction of the relative probabilities for the appearance of the individual phenomena under given conditions. (II.73)

Corresponding to the fact that different individual quantum processes may take place in a given experimental arrangement these relations (between observations obtained under well-defined conditions) are of an inherently statistical character. (II.71)

The very fact that repetition of the same experiment, defined on the lines described, in general yields different recordings pertaining to the object, immediately implies that a comprehensive account of experience in this field must be expressed by statistical laws. (III.4)

The fact that in one and the same well-defined experimental arrangement we generally obtain recordings of different individual processes makes indispensable the recourse to a statistical account of quantum phenomena. (III.25)

These statements indicate a turning away by Bohr from picturesque notions of an inherent random element in nature itself, and the adoption of an essentially pragmatic attitude toward the statistical character of the quantum-mechanical predictions.

It is worth noting that Bohr's notion of complementarity is altogether pragmatic: Ideas should be judged by their utility; physical ideas should be judged by their success in ordering physical experiences, not by the accuracy with which they can be believed to mirror the essence of external reality. The use of complementary ideas in complementary situations is a natural concomitant of pragmatic thinking.

3.6 Spacetime and the Completeness of Quantum Theory

In spite of doubts cast on our intuitive notions of space and time by the theory of relativity, the idea lingers on that persisting physical objects occupy spacetime regions that can be divided into ever finer parts. A basic premise of classical physics is that this classical concept of the spacetime continuum is the appropriate underlying concept for fundamental physical theory.

It is important to recognize that quantum theory has nothing in it that can be regarded as a description of qualities or properties of nature that are located at the point or infinitesimal regions of the spacetime continuum. On one hand, the descriptions of the experimental arrangements and observations are basically operational descriptions of what technicians can see and do. They are not, strictly speaking, descriptions of the external things in themselves. Moreover, they are not descriptions of *microscopic* qualities or properties. On the other hand, the wave functions are merely abstract symbolic devices. They do not describe qualities or properties of nature that are located at points or infinitesimal regions of the spacetime continuum. The abrupt change of a wave function in one region of spacetime when a measurement is performed far away at the same time makes any such interpretation unreasonable. The wave functions of quantum theory are to be interpreted as symbolic devices that scientists use to make predictions about what they will observe under specified conditions. As Bohr says it:

> In the treatment of atomic problems, actual calculations are most conveniently carried out with the help of a Schrödinger state function, from which the statistical laws governing observations attainable under specified conditions can be deduced by definite mathematical operations. It must be recognized, however, that we are here dealing with a purely symbolic procedure the unambiguous physical interpretation of which in the last resort requires a reference to the complete experimental arrangement. (III.5)

> In fact, wave mechanics, just as the matrix theory, represents on this view a symbolic transcription of the problem of motion of classical mechanics adapted to the requirements of quantum theory and only to be interpreted by an explicit use of the quantum postulate. (I.75)

The fact that quantum theory contains nothing that is interpreted as a description of qualities located at points of an externally existing spacetime continuum can be construed as evidence of its incompleteness. However, all we really know about the spacetime continuum is that it is a concept that has been useful for organizing sense experience. Man's effort to comprehend the world in terms of the idea of an external reality inhering in a spacetime continuum reached its culmination in classical field theory. That theory, though satisfactory in the domain of macroscopic phenomena, failed to provide a satisfactory account of the microscopic sources of the field. The bulk of Einstein's scientific life was spent in a frustrated effort to make these ideas work at the microscopic level.[20] The rejection of classical theory in favor of quantum theory represents, in essence, the rejection of the idea that external reality resides in, or inheres in, a spacetime continuum. It signalizes the recognition that "space", like color, lies in the mind of the beholder.

If the classical concept of the spacetime continuum were accepted, then quantum theory could not be considered complete, i.e., if it were accepted that the persisting objects of nature literally reside in a spacetime continuum, with their various parts definitely located in specific regions, then a complete scientific account of atomic phenomena would, by virtue of the natural and normal meanings of these words, in this framework, be required to describe whatever it was that is located at the points or infinitesimal regions of that continuum. Quantum theory does not do this, and hence a claim of completeness would be an abuse of language.

In a pragmatic framework the claim of completeness has a different natural meaning. The natural meaning of the claim that quantum theory provides for a complete scientific account of atomic phenomena is that no theoretical construction can yield experimentally verifiable predictions about atomic phenomena that cannot be extracted from a quantum-theoretical description. This is the practical or pragmatic meaning of scientific completeness in this context.

The second essential ingredient of the Copenhagen interpretation is the claim that quantum theory provides for the complete scientific account of atomic phenomena. During the more than thirty years spanned by his three books[4] Bohr polished and refined his views on this point. His final, and I think best, summary is as follows:

> The element of wholeness, symbolized by the quantum of action and completely foreign to classical physical principles, has . . . the consequence that in the study of quantum processes any experimental inquiry implies an interaction between atomic object and the measuring tools which, although essential for the characterization of the phenomena, evades a separate account if the experiment is to serve its purpose of yielding unambiguous answers to our questions. It is indeed the recognition of this situation which makes recourse to a statistical mode of description imperative as regards the expectations of the occurrence of individual quantum effects in one and the same experimental arrangement. (III.60)

This statement is augmented and clarified by an earlier statement:

> The essentially new feature in the analysis of quantum phenomena is . . . the introduction of a fundamental distinction between the measuring apparatus and the objects under investigation. This is a direct consequence of the

necessity of accounting for the functions of the measuring instruments in purely classical terms, excluding in principle any regard to the quantum of action. On their side, the quantal features of the phenomena are revealed in the information about the atomic objects derived from the observations. While within the scope of classical physics the interaction between the object and apparatus can be neglected or, if necessary, compensated for, in quantum physics this interaction thus forms an inseparable part of the phenomena. Accordingly, the unambiguous account of proper quantum phenomena must, in principle, include a description of all relevant features of the experimental arrangement. (III.3)

The basic point here is that well-defined objective specifications on the entire phenomenon are not restrictive enough to determine uniquely the course of the individual processes, yet no further breakdown is possible because of the inherent wholeness of the process symbolized by the quantum of action.

This way of tracing the need for a statistical account of atomic phenomena back to the element of wholeness symbolized by the quantum of action appears to take one outside the pragmatic framework, since it refers to the measuring device, the atomic object, and their interaction. Also, it is not immediately clear how one is to reconcile the separate identification of these three things with the

impossibility of any sharp separation between the behaviour of the atomic objects and the interaction with the measuring instruments which serve to define the conditions under which the phenomena appear. (II.39)

In this connection it is important to recognize that the "atomic object" and "measuring instruments" are, within the framework of quantum thinking, *idealizations* used by scientists to bring order into man's experience in the realm of atomic phenomena. I develop this point in reference 21. Bohr's words emphasize that these separate idealizations are inseparably linked by quantum thinking in a way that is completely foreign to classical thinking. The idealization "the measuring instrument" is a conceptual entity used in the description of the experimental specifications; the idealization "the atomic object" is a conceptual entity that is represented by the wave function. These are inseparably linked in quantum theory by the fact that the specifications described in terms of the measuring instrument are mapped onto wave functions associated with the atomic object: the atomic object represented by the wave function has no meaning in quantum theory except via its link to experience formulated in terms of specifications that refer to the measuring instruments.

Bohr evidently believed that there is in atomic processes an element of *wholeness*—associated with the quantum of action, and completely foreign to classical physical principles—that curtails the utility of the classical idealizations of the measuring instruments and atomic objects as separate, interacting entities, and that the resulting inseparability of the atomic object from the whole phenomenon renders statistical description unavoidable.

This way of reconciling the pragmatic character of quantum theory with the claim of completeness seems rational and coherent. It is, of course, based on quantum thinking itself and is therefore essentially a self-consistency consideration. The validity of quantum-mechanical thinking as a whole must, of course, be judged on the basis of its success, which includes its coherence and self-consistency.

The question of the completeness of quantum theory was debated by Bohr[22] and Einstein[23]. Einstein's counterarguments come down to the following points: (1) It is not proven that the usual concept of reality is unworkable; (2) quantum theory does not make "intelligible" what is sensorily given; and (3) if there is a more complete thinkable description of nature, then the formulation of the universal laws should involve their use.

Bell's theorem[16] deals a shattering blow to Einstein's position. For it proves that the ordinary concept of reality is incompatible with the statistical predictions of quantum theory. These predictions Einstein was apparently willing to accept. Einstein's whole position rests squarely on the presumption that sense experience can be understood in terms of an idea of some external reality whose spatially separated parts are independent realities, in the sense that they depend on each other only via connections that respect spacetime separation in the usual way: instantaneous connections are excluded. But the existence of such a reality lying behind the world of observed phenomena is precisely what Bell's theorem proves to be impossible.

Einstein's second point, about whether quantum theory makes intelligible what is sensorily given, is taken up in the next section.

Einstein's third point raises two crucial questions. The first is whether a complete description of nature is thinkable. Can human ideas, which are probably limited by the structural form of human brains, and which are presumably geared to the problem of human survival, fully know or comprehend the ultimate essences? And even if they can, what is the role in nature of universal laws? Is all nature ruled by some closed set of mathematical formulas? This might be one possibility. Another, quite compatible with present knowledge, is that certain aspects of nature adhere to closed mathematical forms but that the fullness of nature transcends any such form.

3.7 Quantum Theory and Objective Reality

The Copenhagen interpretation is often criticized on the grounds that it is subjective, i.e., that it deals with the observer's knowledge of things, rather than those things themselves. This charge arises mainly from Heisenberg's frequent use of the words "knowledge" and "observer". Since quantum theory is fundamentally a procedure by which scientists make predictions, it is completely appropriate that it refer to the knowledge of the observer. For human observers play a vital role in setting up experiments and in noting their results.

Heisenberg's wording, interpreted in a superficial way, can be, and has been, the source of considerable confusion. It is therefore perhaps better to speak directly in terms of the concrete social realities, such as dispositions of instruments, etc., in terms of which the preparations, measurements, and results are described. This type of terminology was favored by Bohr, who used the phrase "classical concepts" to signify descriptions in terms of concrete social actualities.

On the other hand, Bohr's terminology, though blatantly objective, raises the question of how quantum theory can be consistently constructed on a foundation that includes concepts that are fundamentally incompatible with the quantum concepts.

Perhaps the most satisfactory term is "specifications". Specifications are what architects and builders, and mechanics and machinists, use to communicate to one another conditions on the concrete social realities or actualities that bind their lives together. It is hard to think of a theoretical concept that could have a more objective meaning. Specifications are described in technical jargon that is an extension of everyday language. This language may incorporate concepts from classical physics. But this fact in no way implies that these concepts are valid beyond the realm in which they are used by the technicians.

In order to objectify as far as possible our descriptions of the specifications on preparations and measurements we can express them in terms of the "objective" quantities of classical physics. The meaning of these "objective" quantities for us is tied to the fact that we conceive of them as the qualities of an external world that exists independently of our perceptions of it. The formulation of the specifications in terms of these classical quantities allows the human observer to be eliminated, superficially at least, from the quantum-theoretical description of nature: the observer need not be explicitly introduced into the description of quantum theory because the connection between his knowledge and these classical quantities is then shifted to other domains of science, such as classical physics, biology, psychology, etc.

But this elimination of the observer is simply a semantic sleight of hand. Since the conceptual structure of classical physics is recognized as fundamentally an invention of the mind that is useful for organizing and codifying experience, the knowledge of the observer emerges, in the end, as the fundamental reality upon which the whole structure rests. The terms "knowledge of the observer", or "classical description", or "specifications" are just different ways of summing up in a single term this entire arrangement of ideas, which follows from the recognition of the limited domain of validity of classical concepts.

Bohr cites certain ideas from biology and psychology as other examples of concepts that work well in certain limited domains. And he notes that there have been repeated attempts to unify all human knowledge on the basis of one or another of these conceptual frameworks.[24] Such attempts are the natural outgrowth of the absolutist viewpoint, which holds that the ideas of man can grasp or know the absolute essences. The pragmatist, regarding human concepts as simply tools for the comprehension of experience, and averring that human ideas, being prisoners in the realm of human experience, can "know" nothing but other human ideas, would not be optimistic about the prospects of complete success in such ventures. For him progress in human understanding would more likely consist of the growth of a web of interwoven complementary understandings of various aspects of the fullness of nature.

Such a view, though withholding the promise for eventual complete illumination regarding the ultimate essence of nature, does offer the prospect that human inquiry can continue indefinitely to yield important new truths. And these can be final in the sense that they grasp or illuminate some aspect of nature as it is revealed to human experience. And the hope can persist that man will perceive ever more clearly, through his growing patchwork of complementary views, the general form of a pervading presence. But this pervading presence cannot be expected or required to be a resident of the three-dimensional space of naive intuition, or to be described fundamentally in terms of quantities associated with points of a four-dimensional spacetime continuum.

3.8 Appendix A. Philosophic Addenda

Several questions of a philosophic nature have been raised by a critic. This appendix contains my replies.

Question 1: How does one reconcile the commitment of James and Bohr to the public character of sense objects with the radical empiricist doctrine that ideas can agree only with other ideas? Russell's *Analysis of Matter* indicates the difficulty in performing this reconciliation.

Reply: Russell's arguments do not confute the ideas of James and Bohr as I have described them. Both of the latter authors would, I think, readily admit that human experiences are probably not the whole of reality but are probably merely a part of the whole that is related to the rest via some sort of causal-type connection. The critical question, however, is not whether there is in fact a world "out there", but rather what the connection is between the world "out there" and our ideas about it.

Russell argues, essentially, that we can make plausible inferences, based on the structure of our experiences, and build up a reasonable idea of the outside world. James would insist that this whole structure is nothing but a structure of abstract ideas built upon our common experiences, and that the transexperiential world that may somehow "cause" our common experiences never enters into this structure at all.

James evidently believes that his idea of a table is similar to yours and mine. In general, different people's ideas about sense objects are not identical, but they are similar enough to form the basis of effective social communication. There exists, in this sense, a realm of public or shared experiences that form the basis of interpersonal communication. This realm constitutes the primary data of science. The aim of science is to construct a framework of ideas that will link these common, or public, or shared, experiences together in ways that reflect various aspects of the empirical connections that exist between them. Thus the whole structure of science is, quite obviously, a structure that is wholly confined to the world of ideas.

Russell would presumably grant this. But he would argue that we can, nonetheless, make plausible inferences about the world based on the structure of experience. Yet his commitment to rationality requires him, I think, to admit that our ideas *might* not be able to fully comprehend the realities that are the causes of our experience. And if the evidence of science indicates that this possibility is the one realized by nature, then I think his rational approach, based on plausible inferences drawn from available evidence, would require him to admit that this possibility has a good "probability" of being correct.

Although the arguments of Russell do not confute the position of James, as I have described it, there is definitely a basic difference in orientation. Russell embarks on a quest for *certainty* about the external world, but settles for an account to which he assigns high "probability". James views the quest for certainty about the external world as totally misdirected. Certainty in such matters is clearly unattainable. The truly rational course of action is to admit at the outset that our aim is to construct a framework of ideas that is useful for the organization of our experience—and for the conduct of our lives—and to put aside the whole vague question about the connection of ideas to nonideas, and the equally vague question about the "probability" that a certain scheme of ideas gives us a true or valid picture of the world itself.

In any case, the claim that we *can* make valid inferences about the world itself acquires credibility only to the extent that a truly adequate picture of the world itself can be constructed. No such picture exists at present. And the difficulties in constructing a scientific view of the world itself are precisely those admitted by Russell himself, namely the incorporation of quantum phenomena and infinitesimal spacetime intervals. It is precisely these difficulties that force us to fall back to the position of James.

In short, the position of Bohr and James, as I have described it, is not a denial of the causal theory of perception. It is simply a recommendation that we view science not as a quest for a metaphysical understanding of that which lies outside the world of ideas, but rather as an invention of the human mind that man constructs for the purpose of incorporating into the world of human ideas abstract structural forms that capture certain aspects of the empirical structure of man's experience. In this undertaking an important class of data are those experiences that are common to different human observers, such as our common or shared experience of the table about which we all sit. The level of experience at which these common experiences are most similar is the level at which a round table is experienced as a round table, not as an oval two-dimensional visual pattern that depends upon where one sits, or a set of tactile sensations that depend on where one's hand rests. In science we need "objective" descriptions of the experienced world. We need descriptions that do not depend on who it is that has the experience. Operational specifications fill this need. They are descriptions of possible human experiences that do not refer specifically to any particular individual. They allow us to create a science that is thoroughly objective, yet securely rooted in the realm of ideas and experience.

Question 2: In your article[21] on the S-matrix interpretation of quantum theory it was admitted that the pragmatic interpretation of quantum theory leaves unanswered deep metaphysical questions about the nature of the world itself. And it was noted that the apparent absence of unanalyzable entities in quantum theory suggests a "web" structure of nature that somewhat resembles the structure proposed by Whitehead. Does the absence of similar remarks in the present work signify a retraction of the earlier views?

Reply: The aim of the present work is to describe the Copenhagen interpretation. More precisely, the aim is to describe this author's understanding of the essential common ground of Bohr and Heisenberg on the question of the interpretation of quantum theory. The author's own views are an elaboration upon his understanding of the Copenhagen interpretation, and are given in the S-matrix article.

3.9 Appendix B. Correspondence with Heisenberg and Rosenfeld

The views that have been put forth as representations of the Copenhagen interpretation differ widely. Thus the question arises whether my description succeeds in capturing the essence of the Copenhagen interpretation as understood by Bohr and Heisenberg. To shed light on this question I inquired of Heisenberg whether the description given in a first version of this paper seemed to him basically in accord with the views of himself and Bohr, or whether it seemed different in any important way.

Heisenberg replied:

> Many thanks for your letter and for your paper on the Copenhagen interpretation. I think that your text is a basically adequate description of the Copenhagen interpretation, and you probably know that Niels Bohr was very interested in the ideas of William James. I would, however, like to mention one point where you seem to describe the Copenhagen interpretation too rigorously. On p. 35 you ask the question "Can any theoretical construction give us testable predictions about physical phenomena that cannot be extracted from a quantum theoretical description?" and you say that according to the Copenhagen interpretation no such construction is possible. I doubt whether this is correct with respect to, for example, biological questions. Logically it may be that the difference between the two statements: "The cell is alive" or "The cell is dead" cannot be replaced by a quantum theoretical statement about the state (certainly a mixture of many states) of the system. The Copenhagen interpretation is independent of the decision of this point. It only states that an addition of parameters in the sense of classical physics would be useless. Besides that it may be a point in the Copenhagen interpretation that its language has a certain degree of vagueness, and I doubt whether it can become clearer by trying to avoid this vagueness.

The paper was revised so as to make it absolutely clear that the claim of completeness of quantum theory refers specifically to atomic phenomena.

Some superfluous material was eliminated, and the present sections 3.5 and 3.7, with their extensive quotations from Bohr, were added. Heisenberg's comments on the revised version were as follows:

> Many thanks for sending me the new version of your paper on the Copenhagen interpretation. It is certainly an improvement that you quote Bohr extensively, and your whole paper has become more compact and more understandable after these changes. There is one problem which I would like to mention, not in order to criticize the wording of your paper, but for inducing you to more investigation of this special point, which is however a very deep and old philosophical problem. When you speak about the ideas (especially in [section 3.4]), you always speak about the human ideas, and the question arises, do these ideas "exist" outside of the human mind or only in the human mind? In other words: Have these ideas existed at the time when no human mind existed in the world?
>
> I am enclosing the English translation of a passage in one of my lectures in which I have tried to describe the philosophy of Plato with regard to this point. The English translation was done by an American philosopher who, as I think, uses the philosophical nomenclature correctly. Perhaps we could connect this Platonic idea with pragmatism by saying: It is "convenient" to consider the ideas as existing even outside of the human mind because otherwise it would be difficult to speak about the world before human minds have existed. But you see at these points we always get easily at the limitation of language, of concepts like "existing", "being", "ideas", etc. I feel that you have still too much confidence in the language, but that you will probably find out yourself.

I replied:

> Regarding the question of nonhuman ideas it seems to me unlikely that human ideas could emerge from a universe devoid of idealike qualities. Thus I am inclined to the view that consciousness in some form must be a fundamental quality of the universe. [However] It is difficult to extract from Bohr's writings any commitment on Platonic ideals. Indeed, Bohr seems to take pains to avoid all ontological commitment: He focuses rather on the question of how we as scientists are to cope with the limited validity of our classical intuitions.
>
> In view of Bohr's reluctance to speculate (in print at least) on the nature of the ultimate essences it has seemed to me that the consideration of these matters should not be considered a proper part of the Copenhagen interpretation. If the Copenhagen interpretation is considered to be an overall world view that coincides with the complete world views of both you and Bohr, then there is danger that the Copenhagen interpretation may not exist; for it is not clear (from your respective writings at least) whether you and Bohr are in complete agreement on all ontological and metaphysical questions. Moreover, in your work *Physics and Philosophy* you discuss many of these deeper philosophical questions, yet have a separate chapter entitled "The Copenhagen Interpretation". This suggests that "The Copenhagen Interpretation" should be interpreted in a restricted way. I have interpreted it to he not the complete overall joint world view of Bohr and yourself, but rather

the essential common ground of you and Bohr on the specific question of how quantum theory should be interpreted.

My practical or pragmatic account of quantum theory was based on the account given in your chapter "The Copenhagen Interpretation". This concrete account jibes completely with the abstract pronouncements of Bohr, as the quotations of Bohr in my [section 3.5] bear witness. Thus I think it correct and proper to regard the pragmatic interpretation of the formalism as an integral part of the Copenhagen interpretation. Similarly, I drew from our conversations at Munich an understanding of your commitment to the position that quantum theory provides for a complete description of atomic phenomena, and this position seems completely in accord with that of Bohr. Thus I think it correct and proper to regard also this position as an essential part of the Copenhagen interpretation. But in view of Bohr's silence on Platonic ideals I would hesitate to include considerations on that question in my account of the "logical essence of the Copenhagen interpretation". This is not meant to suggest that the Copenhagen interpretation bans further search for a comprehensive world view. It indicates only that the Copenhagen interpretation is, in my view, not itself a complete overall world view: It is merely part of an overall world view; the part that establishes the proper perspective on quantum theory. I emphasized in the closing passages of my paper that man's search for a comprehensive world view is not terminated by the Copenhagen interpretation. Rather it is significantly advanced.

Heisenberg replied:

Many thanks for your letter. May I just briefly answer the relevant questions. I agree completely with your view that the Copenhagen interpretation is not itself a complete overall world view. It was never intended to be such a view. I also agree with you that Bohr and I have probably not looked upon the Platonic ideals in exactly the same way, and therefore there is no reason why you should go more into the problems of the Platonic ideals in your paper. Still there is one reservation which I have to make in connection with your paper and which I mentioned in my last letter. I think that you have too much confidence in the possibilities of language. I think that the attitude which is behind the Copenhagen interpretation is not compatible with the philosophy of Wittgenstein in the *Tractatus*. It may be compatible with the philosophy contained in the later papers of Wittgenstein. As you probably know, Bertrand Russell liked the *Tractatus* of Wittgenstein, but disapproved of the later papers, and therefore I could never come to an agreement with Russell on these philosophical questions.

I replied:

Thank you for your very informative letter. I had not previously fully appreciated the point you were making, which as I now understand it, is this: You regard recognition of imperfectability of language to be an important element of the attitude that lies behind the Copenhagen interpretation. This point was not brought into my account of the Copenhagen interpretation, and is indeed somewhat at odds with its avowed aim of clarity . . . [But] scientists must strive for clarity and shared understandings, since without striving even the attainable will not be achieved . . .

> Your words on the matters raised in our correspondence would certainly be extremely valuable to readers of my article. And any paraphrasing I might make would diminish this value. Thus, with your approval, I would like to include the full content of your letters (apart from personal openings and closings) in an appendix to my paper, along with certain connecting excerpts from my own letters. I have enclosed a copy of the proposed appendix, apart from your reply to the present letter.

Heisenberg replied:

> Many thanks for your letter. I agree with your intention to publish my letters in the appendix to your paper.

I inquired also of Rosenfeld, as the close companion and coworker of Bohr, and prime defender of his views, for an opinion of the extent to which my description succeeded in capturing the essence of the Copenhagen interpretation as it was understood by Bohr. Rosenfeld expressed full agreement with my account, and gave hearty approval. He went on to comment on the relationship of Bohr to James. I include his remarks because of their historical interest:

> It may interest you to know that I several times endeavoured to persuade Bohr to make explicit mention of the affinity between his attitude and that of James, but he firmly refused to do so; not because he disagreed, but because he intensely disliked the idea of having a label stuck onto him. Indeed you may have noticed that some philosophers are already busy tracing imaginary influences of various philosophers upon Bohr. With regard to William James, I am quite sure that Bohr only heard of him from his friend, the psychologist Rubin, and from myself in the '30's. He then expressed enthusiastic approval of James' attitude, which he certainly felt akin to his own; but it is a fact—a very significant one, I think—that James and Bohr developed a pragmatic epistemology independently of each other.
>
> It might be advisable to add somewhere in your paper a remark to that effect in order to avoid further misunderstanding. As a matter of fact, I have never myself in the papers I wrote on complementarity brought the pragmatic aspect of Bohr's thinking in explicit relation with James, precisely in order to avoid such misunderstanding.

He went on to say:

> I notice from your further letters with new title pages that you hesitate about the best title for your essay. I have no very strong view about this, but I would incline to prefer your March 31 title ["Quantum Theory, Pragmatism, and the Nature of Spacetime"], the reason being that it does not contain the phrase "Copenhagen interpretation", which we in Copenhagen do not like at all. Indeed, this expression was invented, and is used by people wishing to suggest that there may be other interpretations of the Schrödinger equation, namely their own muddled ones. Moreover, as you yourself point out, the same people apply this designation to the wildest misrepresentations of the situation. Perhaps a way out of this semantic difficulty would be for you

to say, after having pointed out what the difficulty is, that you make use of the phrase "Copenhagen interpretation" in the uniquely defined sense in which it is understood by all physicists who make a correct use of quantum mechanics. Surely, this is a pragmatic definition.

References

1 K. R. Popper and M. Bunge, in *Quantum Theory and Reality*, edited by M. Bunge (Springer, New York, 1967).

2 L. E. Ballentine, *Rev. Mod. Phys.* **42**, 358 (1970).

3 E. Bastin (editor), *Quantum Theory and Beyond* (Cambridge University Press, Cambridge, 1970); L. Rosenfeld, *Suppl. Prog. Theoret. Phys.* (extra number) 222 (1965); J. M. Jauch, E. P. Wigner, and M. M. Yanase, *Nuovo Cimento* **48B**, 144 (1967); L. Rosenfeld, *Nucl. Phys. A* **108**, 241 (1968); A. Loinger, *Nucl. Phys. A* **108**, 245 (1968); A. Fine, *Phys. Rev. D* **2**, 2783 (1970).

4 N. Bohr, *Atomic Theory and the Description of Nature* (Cambridge University Press, Cambridge, 1934); *Atomic Physics and Human Knowledge* (Wiley, New York, 1958); *Essays 1958/1962 on Atomic Physics and Human Knowledge* (Wiley, New York, 1963).

5 W. Heisenberg, *The Physical Principles of the Quantum Theory* (Dover, New York, 1930); in *Niels Bohr and the Development of Physics*, edited by W. Pauli (McGraw-Hill, New York, 1955), p. 12; *Physics and Philosophy* (Harper and Row, New York, 1958); *Daedalus* **87**, 95 (1958).

6 C. F. von Weizsäcker, in *Quantum Theory and Beyond*, edited by E. Bastin (Cambridge University Press, Cambridge, 1970).

7 J. von Neumann, *Mathematical Foundations of Quantum Mechanics* (Princeton University Press, Princeton, 1955).

8 E. Wigner, *Am. J. Phys.* **31**, 6 (1963).

9 G. Ludwig, in *Werner Heisenberg und der Physik unserer Zeit* (Friedrich Vieweg, Braunschweig, 1961).

10 E. Wigner, in *The Scientist Speculates*, edited by I. J. Good (Basic Books, New York, 1962), p. 284.

11 H. Everett III, *Rev. Mod. Phys.* **29**, 454 (1957).

12 J. A. Wheeler, *Rev. Mod. Phys.* **29**, 463 (1957).

13 Bryce DeWitt, *Phys. Today* **23**, 30 (Sept. 1970).

14 L. Rosenfeld, *Nucl. Phys. A* **108**, 241 (1968).

15 D. Bohm, *Phys. Rev.* **85**, 166, 180 (1952).

16 J. S. Bell, *Physics* **1**, 195 (1964) and Varenna Lectures, Preprint TH.1220-CERN, Aug. 1970. See also ref. 21.

17 W. James, *The Meaning of Truth* (University of Michigan, Ann Arbor, 1970). This reference to James does not mean that the ideas reviewed in this section are exactly those of James or wholly those of James. Countless philosophers have said similar things.

18 Ref. 17, p. 239.

19 Ref. 17, p. 217.

20 A. Einstein, in *Albert Einstein, Philosopher-Scientist*, edited by P. A. Schilpp (Tudor, New York, 1951), p. 675.
21 H. P. Stapp, *Phys. Rev. D* **3**, 1303 (1971).
22 N. Bohr, *Phys. Rev.* **48**, 696 (1935); and in ref. 20, p. 201.
23 A. Einstein, in ref. 20, p. 665.
24 For an interesting and very readable account of the four principal conceptual structures that have been advanced as the basis of overall world views see S. C. Pepper, *World Hypothesis* (University of California Press, Berkeley, 1970).

4 Mind, Matter, and Quantum Mechanics

4.1 Introduction

The purpose of this work is to resolve together four basic questions concerning the nature of nature. These questions are: (1) How is mind related to matter? (2) How is quantum theory related to reality? (3) How is relativity theory reconciled globally with that which locally we experience directly, namely the coming of reality into being or existence? (4) How is relativity theory reconciled with the apparent demand of Bell's theorem that what happens in one spacetime region must, in certain situations, depend on decisions made in a spacelike-separated region? These four questions will be discussed in detail later. They are probably the four most fundamental questions in science.

The resolution of these questions proposed here is based on a modified Whitehead–Heisenberg ontology according to which all that exists is created by a sequence of creative acts or events, each of which brings into being one possibility from the multitude created by prior acts. The focus of the present work is on those special creative acts that correspond to conscious experiences, and a testable model of the relationship between conscious experiences and neural events is proposed. This proposed solution of the mind–body problem requires no ad hoc distortion of the laws of physics. Instead, it arises naturally from the simplest way of conceiving a universe in which the laws of relativistic quantum theory hold. The nature of the proposed mind–body connection is in general accord with some ideas advanced by the neurobiologists R. W. Sperry and J. C. Eccles, but is much more specific.

The organization of the paper is as follows. Section 4.2 introduces the mind–body problem through the words of William James, Charles Sherrington, and R. W. Sperry. Section 4.3 gives a brief account of the basic conceptual framework of quantum theory as it relates to the mind–body problem and to the present work. Section 4.4 gives a sharpened version of the author's earlier formulation of Bell's theorem, with a detailed discussion of the key assumption about the effective freedom of the experimenters. These first four sections provide the necessary background for the main body of the work, which is the theory of psychophysical reality presented in section 4.5.

Section 4.5 is divided into 18 subsections. The first 11 describe the basic ontology, which is similar to Whitehead's: reality is created by a sequence of self-determining creative acts; the physical world, as represented by the waveform (i.e., the wave function) of quantum theory, represents tendencies for the creative acts; each creative act is represented in the physical world (as represented in quantum theory) by a collapse of the waveform. Section 4.5.12 shows how this ontology accounts quantitatively for the non-local transfer of information apparently demanded by Bell's theorem.

Section 4.5.13 explains how this theory can be reconciled with the theory of relativity. It is noted that relativity theory and quantum theory are both based on Einstein's conceptualization of physical theory as a structure of mathematical relationships between the elements of Einstein's static realm of readings of devices. The notion of process, i.e., of the ongoing process of the unfolding of nature, has no place in this realm, whose elements have, moreover, an ambivalent status as regards their assignment to the worlds of mind and matter. The fact that the statistical regularities described by relativistic quantum theory can be formulated within the limited framework provided by Einstein's realm of readings does not imply that the full understanding of nature must be formulated in this limited way. Indeed, the unreasonableness of imposing upon process conditions drawn from Einstein's static realm of readings is noted, and the apparent logical inconsistencies that arose in Whitehead's attempt to do this are analyzed. This analysis provides the rational basis for the fundamental assumption made here that the creative acts are arranged in a well-ordered linear sequence. This ordering of the creative acts does not disrupt the Lorentz invariance of the statistical predictions of quantum theory, which arises naturally from general properties of the creative process.

Section 4.5.14 applies the general ontological structure developed in the earlier subsections to the problem of the connection between brains and consciousness. On the basis of the results of recent neurobiological research a model of a system of mutually exclusive self-sustaining patterns of neural excitations is proposed. This primary system is linked to a secondary system, the memory system, which records, by enduring structural changes, images of the self-sustaining patterns that occur in the primary system. Neurological mechanisms are postulated that can, by using the templates stored in memory, activate within the primary system patterns having parts that resemble parts of patterns whose images were previously stored.

The dynamical evolution of the physical brain according to the dynamical laws of quantum theory generates in the conscious brain a superposition of many different mutually exclusive self-sustaining patterns with different statistical weights. The image in physical theory of the conscious act is the act of selecting one of these patterns. The information content of the conscious thought is contained in the self-sustaining pattern of neural excitation that is selected by this conscious act.

Sections 4.5.15 and 4.5.16 describe some ideas of the neurobiologists R. W. Sperry and J. C. Eccles. According to these ideas, consciousness exercises top-level control over the neural excitations of the brain. This feature is incorporated into the present theory in sections 4.5.17 and 4.5.18, where it is specified that the brain functions as a self-programming computer, that the aforementioned mutually exclusive self-sustaining patterns of neural excitations constitute the top-level code, and that each human experience is a conscious act that is represented in the physical world as described by quantum theory by the selection of a top-level code that is functionally equivalent to the experience. Thus, conscious experience, as represented in the physical world described by quantum theory, exercises precisely the top-level control that is consciously experienced. The theory is thus in accord with the main thrust of the ideas of Sperry and Eccles, but is much more detailed and specific, and overcomes the main objections to their ideas, which is the lack of a clear reconciliation with the laws of physics. In the present theory consciousness enters neither as a mere collective action nor as an ad hoc supernatural agent still to be reconciled with the laws of physics. It enters rather as a process actually demanded by the contemporary laws of physics if the physical world represented by the waveform of quantum theory is to be kept in line with the world we experience.

Section 4.6 discusses tests, applications, and implications of the theory. Section 4.7 contrasts the understanding of the mind–matter connection obtained here to the lack of understanding provided by some other ways of interpreting quantum theory.

4.2 Mind and Matter

The idea that nature has two parts, one containing feelings and thoughts, the other material objects in motion, was created in antiquity. Revived in modern times by Descartes, it became the foundation for classical physics. But man, having thus put nature asunder, was then unable to see her whole. The problem was well described by William James:

Everyone admits the entire incommensurability of feeling as such with material motion as such. "A motion became a feeling!"—no phrase that our lips can form is so devoid of apprehensible meaning. Accordingly, even the vaguest of evolutionary enthusiasts, when deliberately comparing material with mental facts, have been as forward as anyone else to emphasize the "chasm" between the inner and outer worlds.

"Can the oscillations of a molecule," says Mr. Spencer, "be represented side by side with a nervous shock [he means a mental shock], and the two recognized as one? No effort enables us to assimilate them. That a unit of feeling has nothing in common with a unit of motion becomes more than ever manifest when we bring the two into juxtaposition."

And again

"Suppose it to have become quite clear that a shock in consciousness and a molecular motion are the subjective and objective faces of the same thing; we continue utterly incapable of uniting the two, so as to conceive that reality of which they are the opposite faces."

In other words, incapable of perceiving in them any common character. So Tyndall, in that lucky paragraph which has been quoted so often that everyone knows it by heart:

"The passage from the physics of the brain to the corresponding facts of consciousness is unthinkable. Granted that a definite thought and a definite molecular action in the brain occur simultaneously; we do not possess the intellectual organ, nor apparently any rudiment of the organ, which would enable us to pass, by a process of reasoning, from one to the other."

Or in this other passage:

"We can trace the development of a nervous system and correlate with it parallel phenomena of sensation and thought. But we soar into a vacuum the moment we seek to comprehend the connection between them ... there is no fusion between the two classes of facts—no motor energy in the intellect of man to carry it without logical rupture from one to the other."[1]

In a similar vein R. W. Sperry writes in 1952:

The comment of Charles Sherrington remains as valid today as when he wrote it more than eighteen years ago: "We have to regard the relation of mind to brain as still not merely unsolved but still devoid of a basis for its very beginning." It is not a solution which we aspire to but only a basis on which to begin.[2]

This aspiration motivates the present work.

The difficulty encountered by the authors quoted above in the task of reconciling the conceptions of mind and matter stems from their tacit acceptance of the conceptualization of matter provided by classical physics, and from the absence of a natural place for thoughts in the physical world as conceived in classical physics. According to classical physics, the physical world consists of particles and fields that evolve in accordance with deterministic laws of motion. Any additional real thing that is related to the physical world must be related to configurations of the particle and field motions. But any such addition is gratuitous: there is no reason for any such addition, and no natural place for it. And the physical world would evolve in the same way with or without it.

The physical configuration associated with a thought may play an important role in the evolution of the physical world. But the integrated holistic experienced thought is conceptually nonidentical to the associated state of motion of the billions of particles. What confers special status to these particular configurations of motions of billions of particles? How are they singled out to be experienced as a whole? And what is the relationship of the experienced content of the integrated thought to the individual motions of the billions of particles?

The theme of this work is that these questions, though unanswerable in any satisfactory way within the conceptualization of the physical world provided by classical physical theory, have a natural answer within the quantum-theoretic conceptualization.

4.3 Quantum Theory and Mind–Matter

Classical physics works well in many situations, but is inadequate for problems involving the atomic or subatomic structure of objects and materials. For problems of this kind one must use quantum theory, which supercedes classical theory in that it reproduces all the experimentally validated predictions of classical theory, and covers the atomic and subatomic domains as well.

The conceptual framework of quantum theory is profoundly different from that of classical physics, and it allows mind and matter to be seen as the natural parts of a single whole. Indeed, the basic change wrought by quantum theory is precisely a transformation of the physical world from a structure lying outside of mind to one that reaches into mind. This metamorphosis is now explained.

The logical structure of quantum theory is closely tied to the way it is used in practice. To use quantum theory a scientist defines a set of operational specifications A on the devices that are going to prepare some system, and a set of operational specifications B on the responses of devices that are going to detect some properties of this system. The specifications A are transformed into a weight function $\rho_A(x, p)$, and the specifications B are transformed into an efficiency function $\rho_B(x', p')$. Quantum-theoretic rules are then used to calculate the propagation function $U_{AB}(x', p'; x, p)$, which transforms the function $\rho_A(x, p)$ from the spacetime location of the preparation to the spacetime location of the detection. Then the probability

$P(B, A)$ that the response will satisfy specifications B if the preparations satisfy specifications A is calculated from the formula[3]

$$P(B, A) = \int dx' \, dp' \, dx \, dp \, \rho_B(x', p') \, U_{AB}(x', p'; x, p) \rho_A(x, p).$$

This formula is identical to the one used for the same purpose in classical statistical mechanics. There the quantity $\rho_A(x, p)$ is the phase-space probability density associated with the initial specification A, and $\rho_B(x', p')$ is the probability that the response of the detectors will meet specifications B if the detected system is characterized by the phase-space point (x', p'). (For an n-particle system x' is a set of $3n$ variables that specifies the positions of the n particles, and p' is a set of $3n$ variables that specifies the momenta of these particles.)

The description given above stresses the close connection between quantum theory and classical statistical mechanics. But important differences also exist. Most important are interference effects, which are exhibited, for example, in the double-slit experiment.

The double-slit experiment is well known: light from a tiny monochromatic source is allowed to pass through a first screen containing two narrow slits and fall onto a second screen. The distribution of light on the second screen is grossly different from the sum of the distributions that would be obtained if each slit were opened separately. This difference is explained quantitatively by assuming that light has a wave structure: the parts of the wave traveling through the two slits can interfere constructively in some areas of the second screen and destructively in other areas to produce the observed interference pattern. But a second aspect of the experiment is the quantization of light: the energy is emitted from the source in discrete units called quanta, which are absorbed as units in tiny regions of the second screen.

The double-slit experiments provide prima facie evidence that light consists of both particles and waves. The idea that the energy is carried by tiny particles that are guided by waves that pass through both slits can account quantitatively for both the quantization and interference effects. This guider-wave idea was studied by de Broglie[4] and successfully completed, in the nonrelativistic approximation, by Bohm[5].

Bohm's model has both waves and particles. The particles are conceptually identical to the point-particles that occur in classical physics. However, the probabilities $P(B, A)$ can be calculated from a knowledge of the waves alone. These probabilities $P(B, A)$ are the only quantities of the theory that can be directly compared to experiment. Thus from a practical point of view the particles are superfluous: they add no content that can be tested or verified, or has any practical use. Indeed, there now exist many variations of Bohm's deterministic model that have superimposed stochastic elements, but that are empirically indistinguishable from Bohm's.[6]

The orthodox interpretation of quantum theory dispenses altogether with these superfluous classical particles. It represents any physical system by a waveform alone. Thus an atom is represented by a stable or quasi-stable waveform. The emission of light from an atom is represented by a change of its waveform to one that represents a less energetic state, accompanied by the creation of a waveform that corresponds to the quantum of light. This latter waveform interacts with the waveform of any atom that lies in the region it transverses to provide a waveform having a part that corresponds to the absorption of the quantum of light by that atom. In this part the waveform representing the atom changes to a form representing a more energetic state, while the waveform representing the light quantum disappears.

This orthodox view rests basically on the fact that the information concerning the amount of energy in the quantum of light can be carried just as well by a wave as by a particle. But the particle concept demands information far beyond that of the magnitude of the quantum of energy. It demands also the specification of an exact spacetime path from the emitting atom to the absorbing atom, and even of exact paths of the particles within these atoms. Most physicists believe that this demand for exact spacetime paths originates in our experience with macroscopic phenomena and classical physics, and need not be met by nature itself in the microscopic domain of atomic and subatomic physics. The observed phenomena are represented far more economically and aesthetically without using the notion of classical particles.

The elimination of classical particles means that the functions $\rho_A(x, p)$ and $\rho_B(x', p')$ cannot be interpreted conceptually in the same way as in classical physics. Indeed this possibility was excluded already by the fact that these functions can become negative, which is not compatible with their classical meanings. However, it is only the probabilities $P(B, A)$ that can be directly compared to experience, and these are guaranteed non-negative by the mathematical structure of the theory. The waveforms, and the essentially equivalent quantities $\rho_A(x, p)$ and $\rho_B(x', p')$, are not given individual or separate meanings in orthodox quantum theory: their meanings arise solely from their roles as parts of the formula for the probabilities $P(B, A)$. These probabilities are, empirically, the probabilities that the observed responses will conform to operational specifications B under operational conditions A. No further meaning is to be ascribed to the symbols occurring in the theory. Thus the physical laws represented by quantum theory are not a set of laws governing an independent entity that exists apart from observations. Rather, they define a mathematical structure of statistical relations among observations. In this sense quantum theory, and the physical world represented by quantum theory, reaches into mind.

Although quantum theory, according to the orthodox view, provides merely a set of mathematical rules for calculating the probabilities $P(B, A)$, rather than a detailed picture of what is actually happening in the external world, it does impose through these rules stringent conditions on the character of the underlying reality. The most interesting and important of these is discussed in the following section.

4.4 Bell's Theorem

Bell's theorem[7] imposes stringent conditions on the nature of reality. It arises from an examination of the statistical predictions of quantum theory in certain particular experimental situations. These situations involve two experimenters who, within the confines of two spacelike-separated spacetime regions, first choose some experimental settings and then observe some experimental results. The theorem shows that it is impossible to reconcile the general validity of the statistical predictions of quantum theory with the idea that the results observed by each experimenter could in principle be independent of the apparently free choice of setting made in the spacelike-separated region by the other experimenter: the general validity of the predictions of quantum theory appears to demand strong nonlocal connections that extend over macroscopic distances.

To obtain this conclusion one may consider the following experiment: suppose a pair of low-energy, spin-$\frac{1}{2}$ particles are allowed to scatter off each other in a small spacetime region that is surrounded by an array of fast electronic detectors. These detectors are arranged to cover almost completely a sphere centered on the scattering region. Only two small holes are left uncovered, and these lie at polar extremities of the sphere. The two particles are detected by the fast electronics upon entering the sphere. Thus if they are not detected shortly afterward by the spherical array, then they have escaped through the two holes and, by virtue of the geometric setup, are traveling on trajectories that will lead one into a Stern–Gerlach device D_1 and the other into a Stern–Gerlach device D_2. The arrival times at D_1 and D_2 are such that the devices D_1 and D_2 are confined, during the passage of the particles through them, to the spacetime regions R_1 and R_2, respectively. The spacetime region R_1 contains also a process that generates from some physical numbers that have been brought into R_1 a "random" number that will be used to select one of several predetermined directions along which the axis of the Stern–Gerlach device will be mechanically aligned. A similar arrangement selects the setting of the axis of the device in R_2. The entire process consisting of the selection of the direction of the axis of the device D_1, the deflection of the particle by the device D_1, the subsequent detection of this particle, and the final recording of the result in some memory bank (or in the brain of a human observer) takes place in the spacetime region R_1. The similar set of processes associated with the device D_2 is confined to the spacetime region R_2.

A Stern–Gerlach device has the property of deflecting the particle by a finite amount in one of two directions: the deflection is either in the direction of the (directed) axis of the device or in the opposite direction. The recorded result tells us which of these two possibilities actually occurred.

If the two particles are of the same kind and their energies are sufficiently small, then the particles will emerge from the scattering in what is called the singlet state. This state is recognized experimentally by the fact that if the directions of two axes of the two devices D_1 and D_2 are identical (in an appropriate frame), then the directions of the deflections in D_1 and D_2 are opposite: if the common direction of the two axes is called "up", then one of the two deflections is "up" and the other is "down".

We come now to the crucial point. Suppose the axis in D_1 is chosen by our procedure to lie in some particular direction d, and that the subsequent deflection in D_1 is then observed to lie in some particular direction–which must be either along d or opposite to d. Suppose this particular direction d is also one of the small set of preassigned directions allowed for D_2. We can arrange that there be a large number of conceivable ways in which the direction d might be chosen for D_2. To be definite, suppose that the physical numbers brought into R_2 include the arrival time of a photon from a distant galaxy, the latest teletype Dow–Jones average, and the temperature at the Chicago airport. A computer in R_2 first picks one of these three numbers "at random" and then computes from it a random number that is used to specify the setting of D_2. We suppose this setting has a good chance to be d.

There are many conceivable ways that the direction d could be selected for D_2. But no matter which of these ways is actually used, the direction of the deflection at D_2 will be the same: it must be opposite to the observed direction of deflection at D_1. That is, given the observed direction of deflection at D_1, the deflection at D_2 must be independent of the particular course of events leading to the choice in region R_2 of d.

This independence of the result at D_2 on the manner in which the direction d of D_2 is chosen suggests that in the analysis of the correlations between the directions of the deflections at D_1 and D_2 it is the directions of D_1 and D_2 that are important, not the manner in which these directions are selected or brought into existence. This suggests that in the analysis of these deflections the directions of D_1 and D_2 can be treated as independent free variables.

These heuristic considerations support the key underlying assumption of Bell's theorem, which is that in the analysis of the correlations between the directions of the deflections in D_1 and D_2 one can consider the choices between the several preassigned directions of the axes of these two devices to be independent free variables. This assumption is not that these choices are literally free, in the sense that they have no causal basis whatever, but merely that they are essentially accidental and can be considered free in the analysis of the correlations in the directions of deflection in this experiment.

Of course, we ordinarily take for granted that variables determined by the whimsical choices of experimenters via processes that are left completely unspecified in the description of the experiment under consideration should be considered free variables. But in the case of Bell's theorem this assumption must be emphasized, for it is the only assumption needed to derive a profound conclusion.

The need to regard these choices as effectively free arises from the need to distinguish cause from effect, and to allow the consideration of alternative possibilities.

Suppose now that the regions R_1 and R_2 are spacelike separated. This means that no information can travel from R_1 to R_2 (or from R_2 to R_1) without traveling either faster than light or backward in time. According to the theory of relativity no signal can travel either faster than light or backward in time. This suggests that the information about the free choice of setting made in each region will be unable to reach the other region, and hence that the result observed in each region should be independent of the free choice of setting made in the other region. The principles of relativity also entail that the "order" in which the two choices of settings are made should have no physical significance: the scenario in which D_1 is fixed "before" D_2 is required to be physically equivalent to the scenario in which D_1 is fixed "after" D_2, since "before" and "after" have no invariant meaning for spacelike-separated events.

The foregoing discussion concerns a single pair of particles. Consider next a set of n such pairs that can be separately analyzed by fast electronics, but that are bunched together so that all n particles going to D_1 arrive essentially together, on the scale of the region R_1, and hence that the setting of D_1 is the same for all of them. The analogous conditions are imposed for D_2 and R_2.

A "set of conceivable results S_i" of the n-particle experiment is represented by a list that specifies for each of the n pairs the directions of the deflections at both D_1 and D_2. For any given number n, any given setting of D_1 and D_2, and any set of conceivable results S_i of the experiment, quantum theory prescribes a probability $P(S_i)$. For any collection C_j of distinct sets of conceivable results S_i the probability that the observed set of results will correspond to some unspecified member of the collection C_j is, of course, the sum of the probabilities $P(S_i)$ of the individual members S_i of the collection. This probability is called the probability $P(C_j)$ associated with the collection.

To derive the desired result, consider two possible settings of D_1 (specified by certain angles $\phi_1 = 0°$ and $\phi_1 = 90°$), and two possible settings of D_2 (specified by the angles $\phi_2 = 0°$ and $\phi_2 = 135°$). Let the four combinations of settings be labeled by the index i ($= 1, 2, 3,$ or 4). Then the following mathematical result holds:[8] For any positive number ϵ, there is an integer n and four collections C_i ($i = 1, 2, 3,$ or 4), one for each of the four combinations of settings of D_1 and D_2, such that the following two properties hold:

1   For each of the four collections C_i ($i = 1, 2, 3,$ or 4), the probability $P(C_i)$ associated with C_i is less than ϵ.
2   For any conceivable combination of four sets S_i ($i = 1, 2, 3,$ or 4), one for each of the four possible combinations of the setting of D_1 and D_2, the requirement that the set of results in each of the two regions R_1 and R_2 be independent of the choice of setting in the other region can be satisfied only if at least one of the four sets of conceivable results S_i belongs to the corresponding collection C_i.

This mathematical fact entails that there is no way to reconcile the validity of the statistical predictions of quantum theory for all four combinations of settings with the requirement that what happens in each region could in principle be independent of the choice of setting made in the other region. For suppose we start with a conceivable set of results S_i for some one of the four combinations of settings. If the above-stated independence property is satisfied, then a change in the setting of D_1 (but not D_2) can give a new set of results in R_1, but will leave unchanged the original set of results in R_2.

Alternatively, a change of the setting of D_2 (but not D_1) can give a new set of results in R_2, but will leave unchanged the original set of results in R_1.

The full set of results for three of the four combinations of settings is thereby fixed. To obtain the results in the fourth case (where both D_1 and D_2 are changed) one can follow up the original change of D_1 by a change of the setting of D_2. Alternatively, one can follow up the original change of D_2 by a change of the setting of D_1. The principles of the theory of relativity assert, as already mentioned, that the order in which choices of settings are made in the two spacelike-separated regions has no physical significance. Thus these two ways of ordering the choices should lead to the same final set of results in the final case, in which D_1 and D_2 are both changed. But this condition fixes uniquely the results in the fourth case to be the combination of the changed set of results in R_1 (obtained from changing D_1 and not D_2) with the changed set of results in R_2 (obtained by changing D_2 and not D_1).

The four sets S_i constructed in this way satisfy the independence property stated in part 2 of the mathematical result stated above. Hence that mathematical result 2 entails that for at least one of the four combinations of settings i the associated set of results S_i lies in the specified set C_i of conceivable results. The probability for this is less than the arbitrarily small positive number ϵ. Thus there is no way in which what happens in each region could be independent of the free choice of setting made in the spacelike-separated region without violating the predictions of quantum theory. Moreover, this violation can be made as large as one likes, by choosing ϵ sufficiently small. And there is no way to re-establish the validity of the quantum predictions by taking a still larger value of n. For, by taking n larger, one can make ϵ still smaller: the magnitude of the violation of the quantum predictions increases beyond any bound as the number n of instances in the sample tends to infinity.

This argument is more intricate than those of Bell and of Clauser and Horne, but the result is much stronger. For there are no assumptions about determinism, hidden variables, or objective reality. The conclusion is simply that there is no way for nature to select results that are compatible with both the predictions of quantum theory and the condition that the results observed in each region be independent of the choice of experiment made in the other region.

The appearance of words like "particle" and "device" in the above arguments does not entail any essential use of the notion of objective reality. The argument can be reformulated purely in terms of the experiences of human observers, as was discussed in detail in reference 9.

Section 4.5 will explain, among other things, how the strong nonlocal connections apparently demanded by Bell's theorem can be understood in a natural way without violating the essential principles of the theory of relativity.

4.5 The Psychophysical Theory

The aim of this section is to set forth a theory of psychophysical phenomena that accords with relativity theory and quantum theory, with some recent ideas from the field of neurobiology, and with certain metaphysical principles I find compelling. The central idea is this: the physical world described by the laws of physics is a structure of tendencies in the world of mind. This general idea is latent in Heisenberg's idea of Potentia,[10] and in von Neumann's description of quantum processes.[11] It has been previously advanced by Whitehead,[12] by myself,[13] and by Wigner.[14] In the following subsections this general idea is developed in detail, with particular attention to relativistic and neurobiological aspects.

4.5.1 Mind: The Creative Process

Mind is identified with the process of creation. Everything that exists is created by this process, which consists of a well-ordered sequence of creative acts called events. Any event is prior to all those that follow it in this sequence, and is subsequent to all those that come before it in this sequence. Each creative act is a grasping, or prehension, of all that has been created by prior acts in a novel but unified way. Whitehead's book *Process and Reality*[12] is essentially an elaboration of roughly this idea.

4.5.2 Necessity and Chance

"Naught happens for nothing, but everything from a ground and of necessity" (Leucippus; see, e.g., Russell[15]). This is the law of necessity. Some writers claim to be comfortable with the idea that there is in nature, at its most basic level, an irreducible element of chance. I, however, find unthinkable the idea that between two possibilities there can be a choice having no basis whatsoever. Chance is an idea useful for dealing with a world partly unknown to us. But it has no rational place among the ultimate constituents of nature.

4.5.3 Necessity and Free Will

Man's free will is no illusion. It constitutes his essence. And it rests upon the law of necessity. Any play of chance would falsify the idea that I, from the ground of my essential nature, make a true choice.

4.5.4 Necessity and Predetermination

The law of necessity entails that the process of creation is internally determined. But it is not externally predetermined.

A system is externally predetermined if its development can in principle be predetermined by first forming outside of itself a representation of the system and its laws of development, and then, by applying these laws to that representation, determining, before the fact, how the system will develop.

A system is internally determined if its development is determined by its internal constitution.

The creative process is internally determined. But owing to its wholeness, neither it nor its laws of development can be represented outside of itself. Hence it is not externally predetermined.

Whitehead's similar formula asserts that the world is internally determined and externally free. Both in principle and in practice the only way to determine precisely how nature will unfold is to let it unfold.

4.5.5 Tendency, Propensity, and Probability

An example of tendency is the tendency for "six" to come up on the throw of a loaded die. The number that actually comes up is determined by unknown factors. But the "loading", combined with the conditions of the throw, and some a priori distributions of unspecified variables, create a tendency (or propensity) for "six" to come up. This idea of tendency or propensity can be made quantitative by associating it with the mathematical theory of probability. Popper has developed this "propensity" interpretation of probability and strongly advocated its use in quantum theory.[16]

4.5.6 Emergent Qualities

Each creative act brings into existence something fundamentally new: it creates a novel "emergent" quality.

4.5.7 Consciousness

At the apex of a hierarchical structure in the decision-making process associated with human brains is a subprocess that enjoys two characteristic properties: a record of its acts is stored in the human memory; and it exercises a partial functional control over both its own development and that of other human biological processes. This subprocess is called human consciousness. It is part of a larger subprocess called consciousness, which includes the conscious processes associated with other creatures.

Consciousness is part of the full creative process. The present work is concerned mainly with human consciousness.

4.5.8 Color

Everything that exists was created by the world process called mind. For example, "greenness" is a collection of emergent qualities that play a prominent part in human consciousness. These qualities came into being during the phase of the creative process associated with the growth of consciousness. Prior to that they did not exist.

4.5.9 Spacetime

Spacetime, like color, is an emergent quality that plays a prominent role in human consciousness, and in a certain theoretical activity within consciousness called physics. The success of physics indicates that the concept of spacetime bears an important relationship to the structural properties of the creative process.

4.5.10 Dynamics

To understand the dynamics of the world process it is helpful to consider first the classical approximation. Suppose the force laws and initial boundary conditions were given. Then Newton's laws would completely determine the development of the system. But what determines the initial conditions? The law of necessity demands that everything be determined by necessity. Hence "free" initial conditions are unacceptable.

Imagine, therefore, that the boundary conditions are set not at some initial time, but gradually by a sequence of acts that imposes a sequence of constraints. After any sequence of acts there remains a collection of possible worlds, some of which will be eliminated by the next act. This elimination is achieved by acting on the existing collection with a "projection operator" in phase space that eliminates some members, but leaves the others untouched. The laws of classical physics are not disturbed by fixing the "boundary conditions" progressively in this way.

An analogous sequence of acts can be defined in quantum physics. Thus the acts that constitute the basic world process are represented in quantum theory by a sequence of projection operators, each of which acts in phase space in such a way as to eliminate certain possibilities, but save others. Each act induces a "collapse" of the waveform, which is discussed in more detail in the following subsection.

4.5.11 Collapse of Waveforms

The observation of a track in a cloud chamber is the observation of a sequence of tiny water droplets. These droplets are formed by the passage of a charged stable or quasi-stable "particle" through the chamber.

According to quantum theory the waveform associated with an electron produced by radioactive decay from a heavy nucleus will propagate away from the original nucleus in all directions and then suddenly collapse to a small region the size of the water droplets when the corresponding track in a cloud chamber is observed. This collapse is completely natural for a probability function, and, correspondingly, there is no tendency or propensity for a quantum to be observed in one place immediately after it is observed in a faraway place.

4.5.12 Explanation of Bell's Nonlocality

The nonlocal connection apparently demanded by Bell's theorem arises only after two systems originally in close communion move apart. In this motion the diverging parts sweep out a V-shaped region in spacetime: the original region of communion R_0 lies at the base of the V, and the two spacelike-separated regions R_1 and R_2 lie at the two upper end points.

This V-shaped region is the spacetime region naturally associated with the nonlocal connection. One can imagine that a huge expanding wall of lead fills up the spacetime region between the two sides of the V, and that two huge sliding walls of lead fill up the spacetime region outside the V. The presence of these leads walls leaves unaffected the quantum correlations and hence presumably also the nonlocal connections demanded by these correlations. On the other hand, the insertion of a weak magnetic field at any place in the V-shaped region generally modifies the quantum correlations, and hence also, presumably, the consequent nonlocal connections.

Bell's nonlocal connection is immediately explained by quantum theory if one accepts that the quantum-theoretic waveforms represent tendencies for the responses of the measuring devices. According to quantum theory, there is a waveform that occupies the V-shaped region described above. Actually this waveform consists of two superimposed parts, each of which covers the V-shaped region. The way in which the total waveform decomposes into these two superimposed parts depends on a direction that can be chosen arbitrarily.

Suppose that in the basic creative process the events corresponding to the detection of the results of the experiment in R_1 occur before or prior to those in R_2. We may then choose the arbitrary direction so that one of the two V-shaped waveforms corresponds to a definite deflection "upward" in R_1 and the other superimposed V-shaped waveform corresponds to a definite deflection "downward" in R_1. The superposition of these two parts of the waveform is the spin-space analog of the full spherical wave the spreads out from a radioactively decaying nucleus. In that case there was a sudden jump to a new waveform when a track in a cloud chamber began to form, and the tendencies for future acts were thus suddenly altered. Correspondingly, there is a sudden shift in the composite V-shaped form when the "up" or "down" deflection is detected in R_1: one of the two superimposed V-shaped parts suddenly disappears, along with tendencies associated with it. Thus when this event in R_1 occurs, the tendencies in R_2 are suddenly changed. The way in which these tendencies are changed depends on how the composite waveform was decomposed into the two parts. But this decomposition depended on the way in which the setting was chosen in R_1. Hence the information about the choice of setting in R_1 is transmitted immediately to R_2 via the sudden change in the tendencies in R_2 associated with the disappearance of one of the two V-shaped waveforms. This accounts for the faster-than-light information transfer.

The above description is not just a pictorial description of how one might imagine the information to be transmitted. It is a representation of the quantitative way quantum theory works: the quantum-theoretic calculations can be carried out by associating the collapse of the waveform with the associated changes in the probabilities $P(B, A)$. Thus quantum theory itself provides an immediate quantitative explanation of the faster-than-light information transfer, once it is admitted that the waveforms represent real tendencies for responses of devices or observers.

Prior to Bell's theorem there was a general reluctance to ascribe any real-tendency interpretation to the waveforms of quantum theory, precisely because this interpretation immediately entails faster-than-light information transfer. However, this objection to the real-tendency interpretation is nullified by Bell's theorem, which apparently shows that faster-than-light (or backward-in-time) information transfer is in any case demanded by the statistical rules themselves, independently of the question of interpretation.

A real-tendency interpretation was suggested by Heisenberg, who asserted that the quantum waveforms represent "tendencies for events and our knowledge of events".[10] To clarify this statement it is necessary to specify the nature of Heisenberg's "events". This is made difficult by the reluctance of members of the Copenhagen school to speak of any reality behind quantum theory. (Any such talk undermines the Copenhagen claim of the completeness of the theory.) Consequently, the "event" associated with, for example, the detection of a particle by a device can be represented in orthodox quantum theory only by a change in a waveform or by a change in our knowledge. But a change in a waveform can represent, again, only a change in "tendencies for events and our knowledge of events". Thus one is trapped in a situation where the "event" dissolves always into further tendencies and there is no final identifiable reality upon which these tendencies can act, other than "our knowledge".

The introduction of real creative acts allows Heisenberg's "events" to be identified with these acts, and the waveforms to be identified as representatives of real tendencies for these acts.

This formulation may be merely a detailed statement of what Heisenberg had in mind but was unable to state without jeopardizing the claim that quantum theory is complete.

4.5.13 Compatibility with Relativity

Two features of the theory outlined above appear to conflict with the theory of relativity. The first is the absolute ordering of the creative acts: this seems contrary to Einstein's principle that the ordering of two spacelike-separated events is defined only relative to some chosen coordinate system. The second is the occurrence of faster-than-light transfer of information: this appears incompatible with Einstein's principle that no signal travels faster than light.

The absolute ordering of the creative acts defines the order in which the parts of reality come into existence. Einstein circumvented this whole question of the order in which things come into existence by creating a new conceptualization of the subject matter of physics.

Einstein approached the problem of space and time by considering observations made by physicists. The observations he considered were primarily of clocks and rulers. His theorizing created a new theoretical realm: Einstein's realm of readings of devices. Each element of this realm is an idealized observation consisting of the readings of a set of idealized devices. These devices include one clock and three rules. The four corresponding readings provide a spacetime coordination of the observation.

Three features of Einstein's realm of readings are important. The first is its static nature: the realm is comprised of a fixed collection of entities, called observations, each of which is represented by a fixed set of numbers.

The concept of process or change does not enter into this theoretical realm. Time is represented exclusively by the set of fixed clock readings.

The second important feature of Einstein's realm of readings is the ambivalent status of these readings as regards their assignment to the worlds of mind and matter. This ambivalence allows the readings to be regarded both as the subjective data with which experimental and theoretical physicists must eventually deal, and also as objective data located in the external physical world.

The third important feature of Einstein's realm of readings is that its elements can be regarded as the appropriate subject matter of physical theory. The idealized readings can be considered to represent the objective data that scientists can collect. Einstein's theorizing effectively redefined theoretical physics to be the attempt to create a mathematical structure of relationships between the elements of the static realm of readings, rather than as an attempt to understand or describe the process by which nature unfolds.

Einstein's realm of readings provides the theoretical foundation for quantum theory, and the aforementioned ambivalent status of readings plays an important role in the Copenhagen interpretation; for it allows these readings, regarded as observations by idealized human observers, to be projected into the physical world to form an objective world of readings of devices. These "readings" constitute objective data that quantum theory seeks to correlate. Their ambivalent status creates the blurring of the distinction between the objective and subjective aspects of observations that was so often stressed by Bohr and Heisenberg.

These authors also argue convincingly that, within the theoretical framework provided by Einstein's realm of readings, quantum theory is in principle complete. But then further fundamental progress demands breaking out of Einstein's realm of readings, and coming finally to grips with the question of the relationship of mind to matter. In doing so there is no reason why something so basic to our intuitive grasp of reality as the notion of process, or the unfolding of nature, should continue to be banned. For this notion was banished in the first place only by Einstein's cleverly contrived realm of readings. Once the notion of process is reinstated the question of the order in which the parts of reality come into existence becomes again meaningful.

If spacetime were some pre-existing structure that is filled up by the advancing creative process, then it might be reasonable to think that the full process of creation consists of many subprocesses acting independently in different spacelike-separated regions. If, on the other hand, spacetime is a structure of relationships that develops during the process of creation itself, then the decomposition of this process into independently acting parts on the basis of spacetime aspects becomes unnatural and subject to possible logical contradictions.

To lay bare the possibility of logical contradiction it is useful to consider the model of process proposed by Whitehead, which incorporates a widespread notion of the demands laid down by the theory of relativity. According to Whitehead the creative process consists of a set of distinct creative acts called actual entities. Relative to any actual entity there is a "given" world of actual entities that are "settled, actual, and already become". This given actual world provides determinate data for the creative act.

Whitehead cites the theory of relativity to justify the notion of "contemporary" actual entities: two actual entities are contemporary when neither belongs to the "given" actual world defined by the other. The references to the theory of relativity make clear that Whitehead intends to allow the idea that each actual entity E is associated with a spacetime region R_E, and that its actual world is composed of actual entities whose spacetime regions intersect the union of the backward light cones of the points of R_E.

This geometric picture accords with the relativistic concept that influences can propagate only into the forward light cone. Two actual entities are contemporary when the spacetime region of neither lies in the backward light cone of the other. Then two contemporary creative acts, though possibly related through their mutual dependence on actual entities that lie in the intersection of their respective backward light cones, would proceed in "causal independence" in the sense that neither depends directly on the other.

When two contemporary entities have well-separated spacetime regions there is little difficulty imagining that each creative act proceeds independently on the basis of the settled data in its own actual world. And if spacetime is a pre-existing continuum that is divided into well-defined cells that can be assigned to separate processes, then again there seems to be no problem with the idea that contemporary processes proceed independently: one can, with a little ingenuity, arrange the cellularization of spacetime so that the process of creation can proceed without being blocked by a situation where neither of two neighboring processes can proceed because the backward light cone of each intersects the cell associated with its neighbor. However, this notion of a preassigned cellularization is altogether alien to the ideas of relativity theory. On the other hand, if process is prior to spacetime in the sense that the spacetime region corresponding to each entity is selected by the creative act itself, then one arrives at a Zeno's paradox type of situation where no creative act can proceed because its data are ill defined, and in particular not settled until the data provided by a possible neighboring act are given. That is, if there is no preassigned cellularization of spacetime, but, on the contrary, each creative act selects its own spacetime region, then the property that contemporary acts proceed in causal independence becomes self-contradictory, because the requirement that the regions associated with two contemporary acts be spacelike-separated contradicts the requirement that the choices of these two regions proceed in causal independence.

Whitehead introduced the notion of causally independent contemporary events with the statement:

> Curiously enough, even at this early stage of metaphysical discussion the influence of "relativity theory" in modern physics is important.[12]

This introduction of causally independent contemporary events is indeed curious from Whitehead's point of view. For his main theme was the organic unity of nature, which is disrupted if the process of creation is allowed to have causally independent parts. Moreover, as just emphasized, the notion of causally independent contemporary events appears to contain a logical contradiction. Thus Whitehead apparently sacrificed the logical and organic coherence of his philosophical system to obtain agreement with what he thought to be the demands of relativity theory.

Relativity theory deals, however, specifically with those parts of our understanding of nature that can be formulated within Einstein's static realm of readings, which is explicitly constructed to have no trace of the idea of process. The empirical fact that some part of our understanding of nature can be formulated in terms of readings alone does not imply that a full understanding can be expressed in this limited way. But if, then, process is reintroduced into our description of nature, it is altogether unreasonable to require it to enjoy the relativistic properties characteristic of the completely alien static realm of readings. For it was precisely the elimination of process from this realm that made meaningless the question of the order of spacelike-separated events. And it was the meaninglessness of this order that then entailed, if causes precede effects, the causal independence of spacelike-separated events.

It is unreasonable to impose upon process relativistic demands drawn from the static realm of readings. However, it is important to reconcile the theory of process with the relativistic features of relativistic quantum theory.

An important point in this connection is that whereas an individual actual process depends on the ordering of the events, the predictions of quantum theory are statistical predictions about ensembles defined by operational specifications on the elements of Einstein's static realm of readings. These operational specifications place no conditions on the order in which spacelike-separated events occur. Thus the tendencies associated with these specifications cannot depend on these orderings. Nor can they depend on any absolute frame of reference. For in this theory spacetime is a purely relational construct: there is no absolute frame of reference. Thus, by virtue of the basic structure of the fundamental process and the logical structure of quantum theory, the predictions of quantum theory can depend neither on any absolute frame of reference nor on the order in which spacelike-separated events occur.

The second apparent conflict with relativity theory is the faster-than-light transfer of information. But this is no conflict at all. What Einstein forbade was faster-than-light signals, where a signal means a controlled transfer of information. The same quantum-theoretic rules that lead to the apparent necessity of faster-than-light information transfer exclude the possibility of faster-than-light signals.[17] This rigorous consequence of the quantum formalism does not necessarily mean that there is no way whatever to transmit a signal faster than light. It does mean that any such signal must involve phenomena not adequately covered by quantum theory.

4.5.14 Brains and Consciousness

Within the framework of contemporary quantum theory one can imagine the ultimate experiments in mind–brain research to be such that every neuron in the brain is wired to an apparatus that will record the times at which it fires, and will also, if instructed, induce a firing of this neuron. Additional microdevices will record the microfields at a fine grid of locations in the brain at a closely spaced sequence of times. The spatial extension of each neuron will be mapped out by techniques that do not perceptibly affect the living brain. Other devices will record the subject's verbal reports regarding his conscious activities.

A possible experimental arrangement will introduce sensory inputs that evoke a conscious choice of motor response. The resulting experimental data will presumably show an initial pattern of neural and field activity that can be associated with the entry of the input information into the brain, followed by a pattern of activity associated with a reorganization of this information, followed, eventually, by a pattern of activity associated with the initiation of the consciously chosen motor response.

I shall assume that the analysis of these data will reveal that the input information is reorganized in a way that allows part of it to be incorporated into a self-sustaining pattern of neural activity that is associated with a conscious thought. The nature of this association will be described in due course.

A simplistic but conceivable way in which certain patterns of neural activity might sustain themselves would work as follows. A set of, for example, 100 neurons would be connected to the rest of the brain so that each combination of ten of them would be associated with a corresponding key neuron: this key neuron would be activated if and only if the associated combination of ten neurons fired, and it would then feed back and cause these ten neurons to fire again. There would be roughly 10^{20} different combinations of ten neurons, and this would entail an equal number of key neurons. But this number 10^{20} is vastly greater than the roughly 10^{10} or 10^{11} neurons in the brain. Thus this model is unsatisfactory.

A more economical arrangement would have the simultaneous firings of any pair (in the set of 100) activate a corresponding key neuron, which would then stimulate this pair to fire again. This would require only $\sim 10^4$ key neurons, but the arrangement would tend to produce a chaotic clamor in which all of the 100 neurons are firing incessantly.

An important feature of the neural structure of brains is the presence of inhibitory neurons.[18] These neurons act to inhibit the firings of the neurons to which they are connected. To get an idea of how a self-sustaining pattern could actually arise in the brain one may consider a set of six neurons arranged in three pairs so that if one member of any one of these three pairs fires then the other member will not fire. This inhibitory structure is superimposed on the previously described structure, which in this case would connect each of the 15 possible pairs that can be formed from the six basic neurons being considered. Thus the firing of any pair would tend to re-excite itself, subject to the overriding inhibitory factor.

This system has altogether eight alternative possible self-sustaining patterns of three activated neurons: one member or the other of each of the original three pairs can be excited. But these eight patterns are mutually exclusive: no two of them can be activated at the same time, due to the inhibitory arrangement.

If one now considers this system (or actually a vastly more complex system based on the same principle of mutually exclusive self-sustaining patterns) to be imbedded in the much larger structure provided by the whole brain, and recalls that the full representation of the brain provided by contemporary physical theory gives merely a representation of tendencies for responses, then the state of the brain, as represented by contemporary physics, will, prior to the excitation of one of these self-sustaining but mutually exclusive patterns, represent only the tendencies for the excitations of the various alternative patterns. The choice of which of these patterns is activated is, according to the contemporary laws of physics, a matter of pure chance.

The basic idea of the present psychophysical theory is to identify the selection of one of these mutually exclusive self-sustaining patterns of neural excitations as the image in the physical world, as represented by quantum theory, of a creative act from the realm of human consciousness.

Conscious acts are associated with memory. Thus the self-sustaining neural pattern associated with the conscious act will presumably serve as a template for the production in the brain of an enduring image of this pattern. Physical mechanisms for the formation of this enduring image are already beginning to be understood,[18] but this detail is not important to the main theme being developed here. What is important is that the enduring image of the neural pattern associated with one conscious act can act as a template in the construction of the neural pattern associated with a later conscious act. Thus the physical representation of a conscious act is the selection of a self-sustaining pattern of neural excitations that can contain various subpatterns that are images of subpatterns of patterns associated with various earlier conscious acts.

This arrangement may appear complicated. However, the "wiring" of human brains is vastly more complex than that of any man-made computer.[18] Hence its capabilities should far surpass that of any such computer. A more detailed specification of the computerlike features of the brain will be given presently, after a discussion of some ideas of neurobiologists interested in the mind–brain connection.

4.5.15 Sperry's Model

Before proceeding to a more detailed development of the general idea out-lined above, it will be useful to review the ideas of Sperry, who describes his interpretation of consciousness as follows:

> The current interpretation of consciousness takes issue with the prevailing view of 20th century science. In the present scheme the author postulates that the conscious phenomena of subjective experience do interact on the brain process exerting an active causal influence. In this view consciousness is conceived to have a directive role in determining the flow pattern of cerebral excitation. It has long been the custom in brain research to dispense with consciousness as just an "inner aspect" of the brain process, or as some kind of parallel passive "epiphenomenon", or "paraphenomena" or other impotent byproduct, or even to regard it as merely an artifact of semantics, a pseudoproblem (Boring 1942; Eccles 1966; Hook 1961).
>
> The present interpretation by contrast would make consciousness an integral part of the brain process itself and an essential constituent of the action. Consciousness in the present scheme is put to work. It is given a use and a reason for being, and for having evolved. On these terms subjective mental phenomena can no longer be written off and ignored in objective explanations and models of cerebral function, and mind and consciousness become reinstated into the domain of science . . .
>
> Compared to the elemental physiological and molecular properties, the conscious properties of the brain processes are more molar and holistic in nature. They encompass and transcend the details of nerve impulse traffic

in the cerebral networks in the same way that the properties of the organism transcend the properties of its cells, or the properties of the molecule transcend the properties of its atomic components, and so on. Just as the holistic properties of the organism have causal effects that determine the course and fate of its constituent cells and molecules, so, in the same way, the conscious properties of cerebral activity are conceived to have analogous causal effects in brain function that control subsets of events in the flow pattern of neural excitation. In this holistic sense the present proposal may be said to place mind over matter, but not as any disembodied or supernatural agent.

When it is inferred that conscious forces shape the flow pattern of cerebral excitation, it is not meant to imply that the properties of consciousness intervene, interfere, or in any way disrupt the physiology of brain cell activation. The accepted biophysical laws for the generation and transmission of nerve impulses are in no way violated. The electrophysiologist, in other words, does not need to worry about any of this, provided he restricts himself to analytic neurophysiology. He docs need to be concerned, however, if he wishes to follow a sensory input to conscious levels and to explain how a sensation or a percept is produced, or how the subsequent volitional response is generated . . .

Although the mental properties in brain activity, as here conceived, do not directly intervene in neuronal physiology, they do supervene. This comes about as a result of a higher level of cerebral interactions that involve integration between large processes and whole patterns of activity. In the dynamics of these higher level interactions the more molar conscious properties are seen to supercede the more elemental physiochemical forces, just as the properties of the molecule supercede nuclear forces in chemical interactions.

To put this another way, the individual nerve impulses and associated elemental excitatory events are obliged to operate within larger circuit-system configurations of which they as individuals are only a part. These larger functional entities have their own dynamics in cerebral activity with their own qualities and properties. They interact causally with one another at their own level as entities. It is the emergent dynamic properties of certain of these higher specialized cerebral processes that we interpret as the substance of consciousness.[19]

The foregoing combines important features of both classic dualistic mentalism and monistic materialism. It is mentalistic in that the contents of subjective mental experience are recognized as important aspects of reality in their own right, not to be identified with neural events as these have heretofore been conceived nor reducible to neural events. Further, the subjective mental properties and phenomena are posited to have a top-level control role as causal determinants (Sperry 1976). On these terms mind moves matter. Not only can subjective mind no longer be ignored in science; it becomes a prime control factor in explanatory models. In former theories of consciousness at all acceptable to science, consciousness has been so defined that the causal march of brain mechanisms would proceed the same, whether it is accompanied by subjective experience or not. This is not the case in the present model.[20]

Sperry draws an analogy between his idea of the connection between consciousness and neural activity and the familiar idea of the connection between an organism and its cellular activity, or the connection between a molecule and its atomic or nuclear activity. These latter connections can be viewed classically as the normal connection of an individual to an environment formed of many individuals. In the classical view this connection can, in principle, be reduced to the causal connection between individuals: the collective action of the many individuals of the environment are simply summed up to give a net environmental effect. It may be possible in some cases to isolate conceptually the causally effective collective qualities, and even to construct theories that deal with these collective qualities as new entities. But according to the classical view these collective features are ultimately reducible simply to the properties of the individuals.

If it is this classical viewpoint that Sperry is adopting, then his causal connections between different hierarchical levels become altogether normal and natural. However, consciousness per se becomes irrelevant to the exercise of causal control by the collective environment. The active causal influence exerted by the environment is nothing more than the net effect of the individuals. The superimposed element called consciousness can remain as epiphenomenal as ever.

Sperry's analogies can be interpreted in a classical manner. Indeed, their clarity and reasonableness arise precisely from this fact. However, he is obviously reaching for much more, namely for the idea that certain collective modes are imbued with a holistic unity that goes beyond the simple idea of a collection of individuals acting in unison by virtue of their mutual interactions. For it is only the introduction of this genuinely holistic feature that would justify the introduction of a new entity, consciousness, that is able to exercise control in its own right. But the classical analogies give no idea of how such a genuinely holistic feature could arise or operate within the bounds of the established laws of physics.

The psychophysical theory developed above shows how quantum theory, interpreted in a most natural way, automatically provides for the emergence of consciousness as a distinctive new entity associated with certain specific collective processes in brains. Moreover, as will be shown in the following sections, this new entity automatically exercises control over neural processes in the brain through the action of the established physical laws of nature, not in spite of them. The theory thus shows how Sperry's general ideas can be rooted in, and in fact emerge naturally from, the quantum-theoretic laws of nature.

4.5.16 Eccles's Model

Taking cognizance of Sperry's ideas, Eccles has proposed a different model of brain dynamics in which consciousness again plays a directive role in the flow of neural excitations.[18] In the Eccles model the self-conscious mind "scans" or "probes" the neurons of a certain portion of the brain, called the liaison brain, which consists of certain modules that are "open" to this scanning operation, and then acts back, feebly, on these neurons to exercise directive control over the overall flow of neural activity. The unity of conscious experience comes from a proposed integrating character of the self-conscious mind. But it is left open how the self-conscious mind is able to organize the information extracted from the numerous open modules and form from it a unified conscious thought, and how this conscious thought produces the integrating action on the neural excitations.

The present theory can be considered a more detailed form of Eccles's general idea. However, that general idea, in the form to which it was carried by Eccles, seems to require the self-conscious mind to have an incredible encyclopedic knowledge of the neural circuitry of the brain, in order to make sense of the firings of the liaison brain and bring about its desired ends by exercising feeble control over selected neurons: the self-conscious mind would have to be a truly godlike entity. Indeed, Eccles speaks of its existence after the death of the brain leaving it with nothing to scan.

Eccles likens the self-conscious mind of his model to "a ghost in the machine". Sperry, on the other hand, emphasizes that in his model mind is not a disembodied or supernatural agent.

This description of the ideas of Sperry and Eccles has prepared the way for the presentation of the final and crucial parts of theory being described here.

4.5.17 Consciousness and Control

The brain is viewed in this theory as a self-programming computer, with the aforementioned mutually exclusive self-sustaining neural patterns acting as the carriers of the top-level codes. Each such code exercises top-level control over lower-level processing centers, which control in turn the bodily functions, and, moreover, construct the new top-level code. This new code is constructed by brain processes acting in accordance with the causal quantum-theoretic laws on localized personal data: the new code is formed by integrating, in accordance with directives from the current top-level code, the information coming from external stimuli with blocks of coding taken from codes previously stored in memory. This causal process of construction necessarily produces, by virtue of the character of the quantum-theoretic laws, not just one single new code, but a superposition of many, each with its own quantum-mechanical weight. The conscious act has as its image in the physical world, as represented by contemporary physical theory, the selection of one of these superposed codes.

The selection will be determined almost completely by the causal quantum-theoretic laws acting on the localized personal data, provided only one of the superposed codes has non-negligible weight. But if several of these codes have appreciable weight, then the global and seemingly statistical element will become important. Thus the selection process has, from the quantum-theoretic viewpoint, both a causal-personal aspect and also a stochastic-nonpersonal aspect.

This model of the connection between mind and matter is in general accord with the ideas of Sperry and Eccles, but is more specific. The conscious act is represented physically by the selection of a new top-level code, which then automatically exercises top-level control over the flow of neural excitations in the brain through the action of the quantum-theoretic laws of nature. The unity of conscious thought comes from the unifying integrative character of the conscious creative act, which selects a single code from among the multitude generated by the causal development prescribed by quantum theory.

4.5.18 Objective Control and Subjective Experience

Every human conscious act is experienced by a human being. Thus it is a human experience. Conversely, each human experience is regarded as a human conscious act.

A familiar example of a (human) conscious act is the act of initiating some motor action, such as raising one's arm. The conscious act of initiating this action is the same as the experience of initiating this action. This conscious act is represented in the physical world described by contemporary physical theory as the selection of a top-level code that initiates this action.

This example is now generalized. It is postulated that each human experience is the human conscious act of initiating those perceptible actions that are initiated by the top-level code whose selection is the physical representation of that conscious act. This postulate ensures the functional identity of each human experience and its representation in the physical world.

This functional identity of human experiences and their representations in the physical world resolves the objections mentioned by William James (see section 4.2) to the classical attempts to understand the connection between mind and matter. Those objections stemmed from the complete dissimilarity of the two ideas: the classical idea of a thought has nothing in common with the classical idea of a collection of particles moving in accordance with Newton's laws. But the conscious act of initiating a perceptible action is closely and naturally related to the selection of the code that initiates this action: the latter is the natural image of the former in the physical world represented by quantum theory.

This way of resolving the mind–body problem is philosophically attractive, and it emerges naturally from quantum theory. It follows from the postulate stated above. But how is this postulate to be reconciled with such familiar experiences as seeing a picture or feeling a pain?

Examining a picture elicits experiences of colors, forms, and textures, and of various related associations. According to the present theory this examination is a process in which, at the physical level, the top-level codes are issuing directives to the lower-level processing centers. These top-level codes are instructing the lower-level centers to form from the incoming stimuli and previously stored code-images new top-level codes that resemble themselves as closely as possible, and that also initiate the storage into memory of themselves, and hence the information of interest that is being recognized. Thus an experience of, for example, noticing that a certain patch in the painting is green is, according to the present theory, a conscious act whose physical representation is the selection of a top-level code that initiates the process of storing this information in memory. More generally, any act or experience of recognition is the conscious act whose physical representation is the selection of a top-level code that initiates the transfer into memory of the information that is recognized. The felt experience of "noting" or "noticing" something is the felt experience of initiating the process of storing in memory the noted information.

The top-level code is closely tied to its own memory structure. This code provides an overall control that can link actions that range over an entire lifetime. To provide efficient top-level control the lower-level centers organize the available information into a simplified schema. It is only the elements of this simplified schematic representation of the body, the external environment, and internal ideas that can be incorporated into the top-level code and its memory structure.

An experience of pain is an experience whose physical representation is the selection of a code that initiates the action of registering in this schema damage to some part of the body. In normal circumstances the construction of this code is performed by the lower-level centers acting under the stimulation of signals from the distressed part. If this stimulation is strong, and the lower-level centers are working normally, then the causal laws of quantum theory will virtually assure the selection of such a code. On the other hand, if these centers are not working normally, or if attention is focused elsewhere, so that the selected top-level code is not the one that would normally be induced by the signals from the distressed parts, then there would be no experience of pain, even if the appropriate stimuli are present.

Conversely, if the normal stimuli for the construction of such "pain" codes by lower-level centers were absent, but the lower-level centers were nevertheless constructing such codes with weight close to unity, then the "pain" codes would almost surely be selected and the pain experienced. This theoretical picture is in general accord with the clinical evidence on pain.[21]

By analysis of this kind each human experience is to be identified with a conscious act of initiating certain perceptible actions, and the representation of this act in the physical world is then to be identified as the selection of a top-level code that initiates these actions. Human consciousness thus becomes represented in the physical world, as described by quantum theory, as an agency that exercises precisely the objective control that is subjectively experienced.

4.6 Tests, Applications, and Implications

Some tests, applications, and implications of the psychophysical theory described above are discussed in the following subsections.

4.6.1 Tests and Applications in Mind–Brain Research

The theory gives definite expectations about what brain research should reveal. It should reveal, first of all, the neural connections required to produce and maintain the mutually exclusive self-sustaining patterns of neural excitations that were hypothesized above. The important inhibitory neurons are already known to exist.[18] The "wiring" needed to achieve the self-sustaining excitations must also be present. Moreover, the whole wiring pattern needed for a computer operation of the kind described must exist. The key features are, first, the "liaison brain" consisting of the collection of neurons in which the top-level program is encoded; second, the mechanisms for producing, elsewhere in the brain, enduring images of these codes; and third, the mechanism by which parts of these enduring images can be used as templates for the construction of parts of new top-level codes.*

---

* Note Added in Preparing the Present Book: In the later versions of the theory the idea that the memory is stored "elsewhere" is replaced by the idea that the memory

The expectations described above refer specifically to the neutral structure of the brain, not to consciousness. According to the present theory, each human experience must be accompanied by the activation, in a human brain, of an associated top-level code. This assertion has some immediate experimental consequences. For example, it is known that the excitation of a single neuron can produce characteristic conscious sensations (e.g., a perceived star[18]). According to the present theory, the felt or perceived sensation can occur only if the excitation of the neuron results in the activation of an associated pattern in the top-level code. Thus if this activation is blocked in any way, then there should be no associated sensation, perception, or experience.

As the experimental techniques of brain research develop it may become possible to identify the separate blocks of coding integrated into the top-level code. The present theory demands that no nuance of significance of meaning can be present in a conscious thought unless the corresponding blocks of coding are present in the associated top-level code. Thus the picture of the mind–brain interaction presented here is not the one in which our intelligence stands outside or above the brain and scans it to pick up enough information to allow it to form its own idea of what is going on in the physical world, and then exert some appropriate control measures to effect its subjective aims. On the contrary, all of the elements of the momentary subjective human intelligence are required to be present in integrated form in the momentary top-level physical code.

These assertions can in principle be tested by comparing the physical structure of the top-level code to the experienced content of the thought.

---

is stored in the facilitated patterns themselves, and that the mechanism of recall and formation of a new top-level instruction is mainly a re-activation of a collection of earlier patterns of neural excitations, which were facilitated in connection with earlier conscious thoughts. The term "top-level code" is dropped because it suggests something static rather than the activated pattern of neural firings that (1) serves as the top-level instruction to the unconscious levels of brain processing, and that (2) by virtue of its active status initiates the physical changes in the structure of the brain that "facilitate" this pattern, making it an easily activated component of future top-level instructions.

4.6.2 Implications in the Domain of Traditional Physical Phenomena

The psychophysical theory developed in this paper deals specifically with the mind–body problem. To first approximation it has no ramifications outside that domain. Indeed, the approximate separability of the mind–body question from the subject matter of classical physics is the basis of that science. Likewise the justification of the pragmatic Copenhagen interpretation of quantum theory rests precisely on the fact that the phenomena traditionally dealt with by quantum theory do not depend on the intricacies of the mind–body connection.

On the other hand, the general theory set forth here has, in principle, profound implications. For if the modified Whitehead–Heisenberg ontology described here is really correct, then the primary task of science is to understand more deeply the general nature of the creative processes: what creative acts other than conscious acts occur, and how are they represented?

The proper course of pursuing these questions is to make specific proposals that have both a rational basis and experimental implications. Work is progressing along these lines and will, I hope, be reported later.

4.7 Comments on Parallelistic Interpretations

The empirical validity of quantum theory shows that its mathematical structure corresponds in some way to reality. In fact, the waveforms themselves exhibit an organic unity that gives them an aura of realness not exhibited by their counterparts in classical statistical mechanics. For example, if the detection device is characterized by Ψ_B, and Ψ_B equals Ψ_A, then the probability of detection is unity: the particle is definitely detected. But any change of Ψ_A diminishes the probability of detection. And any change in Ψ_B diminishes the probability of detection. Thus the waveform Ψ_A acts, in this connection, like an organic whole, which is grasped as a whole by the detection device. Its behavior is qualitatively different from what one would expect from a representation of a collection of different independent elements. For if Ψ_A represented a collection of nonidentical elements, some change in Ψ_B, initially equal to Ψ_A, should increase $P(B, A)$.

This characteristic aspect of wholeness in the behavior of the waveform has led many physicists to the idea that the waveform should be considered not merely a calculation tool, but rather a representation of some real aspect of nature itself. According to this view in its extreme form the entire physical world should be represented by a single waveform Ψ. Then the Cartesian dualism familiar from classical physics can be carried over virtually unchanged into quantum theory.

This parallelistic viewpoint was apparently adopted by von Neumann, who says:

> First, it is inherently entirely correct that the measurement or the related process of subjective perception is a new entity relative to the physical environment and is not reducible to the latter. Indeed, the subjective perception leads us into the intellectual inner life of the individual, which is extra-observational by its very nature (since it must be taken for granted by any conceivable observation or experiment). Nevertheless, it is a fundamental requirement of the scientific viewpoint—the so-called principle of the psychoparallelism—that it must be possible so to describe the extra-physical process of the subjective perception as if it were in reality in the physical world—i.e., to assign to its parts equivalent physical processes in the objective environment, in ordinary space.[11]

He also says that

> we must always divide the world into two parts, the one being the observed system, the other the observer. In the former we can follow up all physical processes (in principle at least) arbitrarily precisely. In the latter this is meaningless. The boundary between the two is arbitrary to a very large extent ... That this boundary can be pushed arbitrarily deeply into the interior of the body is the content of the principle of psychophysical parallelism—but this does not change the fact that in each method of description the boundary must be put somewhere, if the method is not to proceed vacuously, i.e., if a comparison with experiment is to be possible. Indeed, experience only makes statements of this type: an observer has made a certain (subjective) observation; and never any like this: a physical quantity has a certain value.
>
> Now quantum mechanics describes the events which occur in the observed portion of the physical world, so long as they do not interact with the observing portion, with the aid of process 2, but as soon as such an interaction occurs, i.e., a measurement, it requires the application of process 1.[11]

(Process 2 is causal development according to the Schrödinger equation, whereas process 1 is an abrupt stochastic change associated with observation or measurement.)

Von Neumann's approach is dualistic and parallelistic: he says that the subjective process can be described "as if" it were in reality in the physical world. He also claims that the boundary between the parts of the world treated as the observed system and the observer, respectively, is arbitrary to a large extent. This evidently means that the abrupt change associated with the process of observation or measurement is not a real process, but merely an artifact of man's theorizing about nature, dependent upon where he places an imaginary cut.

If the abrupt changes called process 1 are not real, as von Neumann's words suggest, then the "real" physical world represented by the waveform

Ψ must develop always causally in accordance with process 2. This leads to odd conclusions. For example, a person looking at a digital clock that is stopped at some time by a radioactive decay would, insofar as his representation in the physical world is concerned, be split into a sequence of copies of himself, each corresponding to a different perceived reading of the clock. More generally the physical world as represented by Ψ will be continually splitting into parts that represent the different perceptual possibilities of all human observers: the one "absolute" real world represented by Ψ will be splitting into parts representing myriads of personal real worlds.[9]

Some physicists unflinchingly accept this "many-worlds" view.[22] Von Neumann himself left this implicit conclusion unstated. Most physicists adopt, when pressed, the agnostic practical position represented by the pragmatic Copenhagen interpretation.

Within the general framework provided by von Neumann's interpretation of quantum theory each creative act of the theory developed here would be represented by a type 1 process.*

The present model is similar in a sense to Bohm's point-particle model, except that the role played by his point-particles is transferred in the present theory to mind. In Bohm's model the waveform Ψ is real, but the positions of his point-particles would determine which of the mutually exclusive self-sustaining patterns of neural excitations is "selected". Being "selected" means that this pattern will be subjectively experienced or felt, whereas the other patterns will not be felt. But this presents a puzzle: why should the presence of these point-particles endow with feeling the particular part of the waveform Ψ in which they lie? For the other parts of the wave function Ψ are equally real, and particles seem, if anything, even less akin to consciousness than waves, which are more holistic.

In the present model the selection of which code is experienced is controlled not by the presence of classical particles but by a fundamentally holistic creative process. This opens the way for some rational understanding of the connection between mind and matter. The dualistic/parallelistic real-particle interpretation of quantum theory makes that connection even more mysterious than ever.

The dualistic/parallelistic many-worlds interpretation likewise provides no possibility for mind to enter in any significant way into the unfolding of physical reality. This way of separating physical science from the larger

---

* Note Added in Preparing the Present Book: This is now interpreted as an actual process in nature itself, occurring when well-separated branches emerge at the macroscopic level.

questions of human existence may appear desirable to some. But in the end it is unacceptable.

4.8 Summary and Conclusions

Four fundamental problems were briefly described in the introduction. The first is the problem of mind and matter, which is the problem of conceiving a reality in which the mental events we experience are related naturally to the physical world represented in physical theory. Historically, the difficulty has been that the physical world represented in classical physical theory consists of tiny localized particles in motion (or perhaps localized field amplitudes) that move in accordance with mathematical laws. This picture gives no clue as to what combinations of motions should correspond to a mental event, or why such events should exist at all. For there is absolutely no place in classical physics for mental events and no need for them; no role for them. And these events seem completely incommensurate with the objects that occur in the theory. Moreover, the feeling of power or efficacy associated with subjective conscious acts must, in this picture, be regarded as completely illusory: consciousness can enter the world only as a passive spectator. Yet this feeling of power pervades our conscious experience; it cannot be simply dismissed as sheer illusion without some explanation or evidence.

The decisive break in the problem of mind and matter was the advent of quantum theory, which showed that the laws of classical physics were not valid, and, moreover, that the simple picture of the physical world provided by classical physics was neither accurate nor adequate. However, quantum theory, in its orthodox interpretation, does not resolve the problem of mind and matter. It circumvents the problem by declining to give any picture at all of the physical world, except the vague one that dimly emerges from the set of statistical rules it provides.

This omission constitutes the second fundamental problem—the problem of quantum theory and reality. The problem here is to formulate a conception of the reality that lies behind the statistical rules of quantum theory.

There are three principal contenders: the many-worlds interpretation, the real-particle interpretation, and the real-tendency interpretation. The first two suffer from a profusion of superfluous entities. Moreover, they provide no basis for the resolution of the problem of mind and matter. In particular, the many-worlds interpretation requires each perceptible world to develop into a multitude of real worlds only one of which we can actually perceive. And consciousness is again, as in classical physics, merely a passive spectator.

The real-particle interpretation, on the other hand, superimposes upon orthodox quantum theory the real particles of classical physics. This wedding is unnatural, and the superimposed real particles are superfluous in the sense that they add nothing that is empirically testable to quantum theory. Their function is merely to single out from the many real worlds of the many-worlds interpretation one single world, which is then identified as the only one that is experienced. This identification does not eliminate or reduce the profusion of worlds generated in the many-worlds interpretation: all of these many worlds are still present in nature, according to this theory. But they are not experienced. Why experience should be associated only with the particular world picked out by the classical particles is not explained. Hence the mind–matter problem is, if anything, magnified.

The real-tendency interpretation was promulgated by Heisenberg, but it seems to conflict with the dogmas of the theory of relativity. However, relativity theory itself is surrounded by long-standing controversies regarding the question of how it should be reconciled globally with that which locally we experience directly, namely the coming of reality into being or existence. This problem of reconciling relativity theory and "process" is the third fundamental problem mentioned in the introduction.

This relativity problem is resolved here by recognizing that Einstein's conception of physical theory identifies it with the construction of mathematical laws that relate various elements from his static realm of readings of devices. This conception eliminates from physical theory, ab initio, the consideration of the process whereby reality comes into being or existence. The ideas and dogmas of the theory of relativity apply naturally only to those aspects of our understanding of nature that can be formulated within Einstein's realm of readings. These aspects are precisely those represented by contemporary physical theory, namely relativistic quantum theory and classical relativity theory. The dogmas of relativity theory cannot be expected to apply to the consideration of the dynamical process by which reality actually unfolds.

This resolution of the relativity problem allows Heisenberg's real tendency interpretation to be formulated in a clear and concrete form. In line with the ideas of Whitehead, reality is conceived to be created by a sequence of creative acts. The quantum-theoretic statistical rules become a reflection of real tendencies induced by the structure of the creative process.

This way of resolving the final three problems mentioned in the introduction leads naturally to a resolution also of the first problem, which is the problem of mind and matter. Starting from the commonly accepted idea that the brain functions as a computer, the present theory identifies each conscious experience with a creative act whose representation in the physical world is the selection of one top-level code from among the multitude automatically generated by the dynamical laws of quantum theory. It is postulated that each conscious experience is the experience of initiating processes that tend to produce certain perceptible changes in the personal reality schema (which consists of the body schema, the external reality schema, and the internal idea schema) and that the representation of this conscious act in the physical world of quantum theory is the selection of the top-level code that initiates the processes that tend to produce these same perceptible changes. Thus the conscious act is functionally equivalent at the level of perceptible changes to its image in the physical world represented by quantum theory. Then the feeling of power of efficacy that pervades the conscious act is no illusion: it correctly represents the functional efficacy of the conscious creative act both in the world of conscious experience and in the physical world represented by quantum theory.

References

1 W. James, *The Principles of Psychology* (Dover, New York, 1950), vol. 1, p. 146.
2 R. W. Sperry, *Am. Sci.* **40**, 291–312 (1952), p. 296.
3 D. Iagolnitzer, *The S-Matrix* (North-Holland, Amsterdam, 1978), p. 61; E. Wigner, *Phys. Rev.* **40**, 749 (1932).
4 L. de Broglie, *An Introduction to the Study of Wave Mechanics* (Dutton, New York, 1930).
5 D. Bohm, *Phys. Rev.* **85**, 166 (1952).
6 E. Nelson, *Phys. Rev.* **150**, 1079 (1966); M. Jammer, *The Philosophy of Quantum Mechanics* (Wiley, New York, 1974); M. Davidson, *Lett. Math. Phys.* **3**, 271 (1979).
7 J. S. Bell, *Physics* **1**, 195 (1964); J. F. Clauser and A. Shimony, *Rep. Prog. Phys.* **41**, 1881 (1978).
8 H. P. Stapp, *Found. Phys.* **9**, 1 (1979); *Epist. Lett.* 36 (1979).
9 H. P. Stapp, *Found. Phys.* **10**, 767 (1980).
10 W. Heisenberg, *Physics and Philosophy* (Harper, New York, 1958).
11 J. von Neumann, *Mathematical Foundations of Quantum Mechanics* (Princeton University Press, Princeton, 1955), chap. 4.
12 A. N. Whitehead, *Process and Reality* (Free Press, New York, 1969).
13 H. P. Stapp, Mind, Matter, and Quantum Mechanics (Zurich, 1959); Mind, Matter, and Quantum Mechanics (University of Nevada Lecture, 1968).
14 E. Wigner, in *The Scientist Speculates*, edited by I. J. Good (The Windmill Press, Kingswood, Surrey, England, 1962).
15 B. Russell, *A History of Western Philosophy* (Simon and Schuster, New York, 1945), p. 66.

16 K. R. Popper, in *Quantum Theory and Reality*, edited by M. Bunge (Springer, Berlin Heidelberg, 1967).
17 W. Heisenberg, *The Physical Principles of the Quantum Theory* (Dover, New York, 1930), p. 39; D. Bohm, *Quantum Theory* (Prentice-Hall, New York, 1951), pp. 618–619; P. Eberhard, *Nuovo Cimento* **46B**, 392 (1978).
18 J. C. Eccles (and K. R. Popper), *The Self and Its Brain* (Springer, Berlin Heidelberg, 1978).
19 R. W. Sperry, *Psychol. Rev.* **76**, 532–536 (1969).
20 R. W. Sperry, in *Brain, Behavior and Evolution*, edited by D. A. Oakley and H. C. Plotkin (Psychology in Progress Series, Methuen, New York, 1979).
21 J. A. M. Frederiks, Disorders of the Body Schema, in *Handbook of Clinical Neurology*, vol. 4, edited by P. J. Vinken and G. W. Bruyn (North-Holland, Amsterdam, and Wiley Interscience, New York), p. 207; R. Melzack, *Perception and Its Disorders* (Research Pub. Assoc. for Research in Nerv. and Ment. Diseases), p. 272.
22 H. Everett III, *Rev. Mod. Phys.* **29**, 454 (1957); and in *The Many-Worlds Interpretation of Quantum Mechanics*, edited by B. DeWitt and N. Graham (Princeton University Press, Princeton, 1973).

5 A Quantum Theory of the Mind–Brain Interface

5.1 The Origin of the Problem: Classical Mechanics

Advances in science often unify conceptually things previously thought to be unconnected. Thus Newtonian mechanics unified our understanding of stellar and terrestial motions, and Maxwell's theory unified our understanding of electromagnetic phenomena and light. Einstein's special theory of relativity unified our concepts of space and time, and his general theory unified our conceptions of spacetime and gravity. My thesis here is that the integration of consciousness into science requires considering together two outstanding fundamental problems in contemporary science, namely the problem of the connection between mind and brain, and the problem of measurement in quantum theory.

Each of these problems concerns the interface between two domains of phenomena that are currently described by using different conceptual systems: mind and brain are described in psychological and physical terms, respectively, whereas the measurement problem in quantum theory is to reconcile the concepts of classical physics that are used to describe the world of visible objects with the concepts of quantum theory that are used to describe the world of atomic processes. In each case the problem of constructing a coherent overarching conceptualization appears to be so intractable that many scientists have judged the problem to be a pseudoproblem not suited to scientific study. However, technological advances are now providing data that bear increasingly on the interfaces between the domains that had heretofore been empirically separate. Given these new data, and the prospect of more to come, science can now profitably take up the challenge of providing a conceptual framework that unifies the mental, physical, classical, and quantal aspects of nature.

William James highlighted the seemingly intractable character of the mind–brain problem with the following two quotations:[1]

> Suppose it to have become quite clear that a shock in consciousness and a molecular motion are the subjective and objective faces of the same thing;

> we continue utterly incapable of uniting the two so as to conceive the reality
> of which they are the two faces. (Spencer)

and

> The passage from the physics of the brain to the corresponding facts of
> consciousness is unthinkable. Granted that a definite thought and a definite
> molecular action in the brain occur simultaneously; we do not possess the
> intellectual organ, nor apparently any rudiment of the organ, which would
> allow us to pass, by a process of reasoning, from one to the other. (Tyndall)

In commenting on this issue James clearly recognized that the problem
was with the concepts of classical physics. Referring to the scientists who
would one day illuminate the problem he said:

> The necessities of the case will make them "metaphysical". Meanwhile the
> best way in which we can facilitate their advent is to understand how great
> is the darkness in which we grope, and never forget that the natural-science
> assumptions with which we started are provisional and revisable things.[2]

James evidently foresaw, on the basis of considerations of the mind–
brain problem, the eventual dislodgement of classical mechanics from the
position it held during his day. We now know that classical mechanics fails
at the atomic level: it has been superseded by quantum mechanics.

That classical mechanics is not capable of integrating consciousness
into science is manifest. Classical physics is an expression of Descartes's
idea that nature is divided into two logically unrelated and noninteracting
parts: mind and matter. However, the integration of consciousness into
science requires, instead, a logical framework in which these two aspects of
nature are linked in ways that can account for both the observed influence of
brain processes on mental processes, and the apparent influence of mental
processes on brain processes.

Brain process depends in a sensitive way upon atomic processes. Hence
a quantum-mechanical treatment is mandated in principle. However, the
brain has a hierarchical structure, with larger structures being built from
smaller ones, and as one moves to higher levels the concepts of classical
physics seem to work increasingly well. Since consciousness appears to be
a high-level process one might think that it should be comprehended within
the conceptual framework of classical physics. In support of this idea some
scientists have noted that, even in nonbiological systems, as one moves
to higher levels of organization new structures often emerge that exercise
effective control over lower-level processes. Thus it is argued that just as a
"vortex" can, within the conceptual framework of classical physics, emerge
as an entity that controls the motions of the molecules from which it is
built, so might there emerge, from a stratum of brain activities completely

compatible with the concepts of classical physics, a "consciousness" that controls lower-level brain processes.

There is, however, an essential conceptual difference between consciousness and a system such as a vortex that is compatible with the concepts of classical physics. The essential characteristic of consciousness is that it is felt: it is felt experience; felt awareness. Any system that is compatible with the concepts of classical physics can be described, insofar as its physical behavior is concerned, as composed of the physical elements provided by classical physics, such as atoms, and electromagnetic fields. However, the description in terms of these elements does not, by itself, specify whether the system has an appended experiential aspect—a feel. Nature may elect to add feel, but the classical physicists can consider the purely physical version without any added quality of feel, and this latter version behaves, according to the precepts of classical physics, in exactly the same way as the one with feel. Thus within the framework of classical physics feel is, per se, nonefficacious: it has no effect on the physical world.

This problem has been clearly understood for hundreds of years, and is the core of the mind–brain problem.

It is only recently that the brain sciences have amassed enough data to make feasible a serious effort to understand the dynamics of the mind–brain connection within the framework of the basic laws of physics. An adequate classical-physics treatment of the mind–brain problem is not possible, for the reason discussed above. On the other hand, the application of quantum mechanics appears to be blocked by three major technical problems.

The first problem, which has already been mentioned, is that quantum theory is primarily a theory of atomic processes, whereas consciousness appears to be connected with macroscopic brain activities, and macroscopic processes are well described by classical physics.

The second problem is that, owing to a failure of an essential condition of isolation, quantum theory, as developed for the study of atomic processes, does not apply to biological systems, such as brains.

The third problem is that the orthodox Copenhagen interpretation of quantum theory instructs us to regard the quantum formalism as merely a set of rules for calculating expectations about our observations, not as a description, or picture, of physical reality itself. However, without a description of physical reality consciousness becomes a puzzle within an enigma.

Any acceptable quantum-mechanical treatment of the connection between mind and brain must resolve these three major technical problems. In the treatment to be described here the resolution of the third problem resolves automatically also the other two.

5.2 A Quantum Ontology

The mathematical concepts in quantum theory are fundamentally different from those of classical physics. This difference makes it difficult to form a unified conception of nature. The Copenhagen strategy for circumventing these conceptual difficulties, by settling for a set of computational rules connecting human observations, rather than striving to comprehend the nature of the underlying reality, was strongly opposed by Einstein, Schrödinger, and many other principal contributors to the development of quantum theory. However, those critics were unable to put forth any alternative proposals. Eventually Werner Heisenberg, one of the chief architects and strongest defenders of the Copenhagen interpretation, did try to form a coherent picture of what is actually happening.[3]

In Heisenberg's picture, which is the one informally adopted by most practicing quantum physicists, the classical world of material particles, evolving in accordance with local deterministic mathematical laws, is replaced by the Heisenberg state of the universe. This state can be pictured as a complicated wave, which, like its classical counterpart, evolves in accordance with local deterministic laws of motion. However, this Heisenberg state represents not the actual physical universe itself, in the normal sense, but merely a set of "objective tendencies", or "propensities", connected to an impending *actual event*. The connection is this: for each of the alternative possible forms that this impending event might take, the Heisenberg state specifies a propensity, or tendency, for the event to take that form. The choice between these alternative possible forms is asserted to be governed by "pure chance", weighted by these propensities.

The actual event itself is simply an abrupt change in the Heisenberg state: it is sometimes called "the collapse of the wave function". The new state describes the tendencies associated with the *next* actual event. This leads to an alternating succession of states and events, in which the state at each stage describes the propensities associated with the event that follows it. In this way the universe becomes controlled in part by strictly deterministic mathematical laws, and in part by mathematically defined "pure chance".

The actual events become, in Heisenberg's ontology, the fundamental entities from which the evolving universe is built. The properties of these actual events are determined by the quantum formalism. These properties are remarkable: they lead to a quantum world profoundly different from the one pictured in classical physics.

Each Heisenberg actual event has both local and global aspects. Locally, each such event acts over a *macroscopic* domain in an *integrative* fashion: it actualizes, *as a unit*, some integrated high-level action or activity, such as the firing of a Geiger counter. This essential quality of the actual event to *grasp as a unit*, and actualize as a whole, an entire high-level pattern of activity injects into the quantum universe an *integrative* aspect wholly lacking in the classical conception of nature. This fundamentally integrative action of the Heisenberg actual event is the foundation of the quantum theory of consciousness developed here.

Each actual event has also a global or universal aspect: its action is not wholly confined to any local region, but extends to distant parts of the universe. These two intertwined aspects arise from the fact that the Heisenberg actual event is represented within the quantum formalism by the change induced in the Heisenberg state of the universe by the action upon it of a localized operator. This change in the state of the universe, although induced by the action of a localized operator, produces a *global* change in the tendencies for the next actual event. Thus each actual event is a global change in the tendencies for the next actual event.

By introducing in this way a quantum *ontology*, and thus departing from the purely epistemological stance of the strictly orthodox Copenhagen interpretation, one can remove the subjective human observer from the quantum description of the physical world and speak directly about the actual dispositions of the measuring devices, rather than the knowledge of the observer. Thus the moon can be said to be "really there" even when nobody is looking. And Schrödinger's cat is, *actually*, either dead or alive. More importantly, the degrees of freedom of a biological system that correspond to its *macroscopic* features can be considered to be highly constrained, and to specify a classical framework, or matrix, within which one can consider the atomic processes that are essential to its functioning.

This useful ontology has two defects. The first is its runaway ontology: the supposedly actual things to which the tendencies refer consist only of shifts in tendencies for future actual things, which consist, in turn, only of shifts in tendencies for still more distantly future things, and so on ad infinitum: each actuality is defined only in terms of possible future ones, in a sequence that never ends.

The second defect is the omission from the description of nature of the one thing really known to exist: human thought.

These two difficulties fit hand-in-glove: the first is that some authentic actual things are needed to break the infinite regress; the second is that some authentic actual things have been left out.

These considerations motivate the first basic proposal of this work, which is to attach to each Heisenberg actual event an experiential aspect. The latter is called the *feel* of this event, and it can be considered to be the aspect of the actual event that gives it its status as an intrinsic actuality.

The central question then becomes: What principle determines the structure of the feel of an actual event? More narrowly: How is the structure of human experience connected to the structure of human brain processes?

The answer, according to the present theory, is this: Each human experience has a compositional structure that is isomorphic to the compositional structure of the actual brain event of which it is the feel.

To understand the nature of these two compositional structures one must look closely at brain processes and psychological processes. We begin by giving a general overview of the former.

5.3 The Functioning of the Brain: An Overview

The primary function of the brain is to gather information about both its environment and the body, to formulate possible plans of action, to choose a *single* plan of action, and to oversee the execution of that plan. Various patterns of neural excitation become activated in the course of these activities. These patterns must presumably represent, among other things, the information that needs to be processed, such as the sensed state of the body and the environment, and the programs for coordinated motor action.

Gerald Edelman has given a scientifically based account of how the brain could have: (1) evolved under natural selection; (2) developed during its individual growth; and (3) become conditioned by its individual history, in such a way as to allow those features that need to be processed to become represented by patterns of neural excitations.[4] One key ingredient is the idea of the *facilitation* of such patterns by physical changes at the synaptic junctions. This process permits certain recurring patterns of excitations in the cerebral cortex that are originally weakly activated by a particular neural activity to become strongly and selectively activated by that activity. Facilitation also permits *association*, whereby the excitation of parts of a facilitated pattern activates, under certain conditions, the rest of that pattern. This association process provides a neural mechanism for retrieval of memories.

To do its job the brain must evidently possess a representation of the body and its environment. I call this representation the body–world schema. A lizard, or a frog, as it watches a moving insect, is, by its attention, continually updating parts of its body–world schema. Quite generally, a basic element of brain operation is the periodic updating, by attention to particular details, of parts of the body–world schema.

When I choose to raise my arm I do not consciously instruct each muscle. I mentally raise my arm to its intended place, and unconscious processes execute the implied instruction. Thus we evidently possess a *"projected body–world schema"* whose content is akin to that of the *"current body–world schema"*, but which specifies a *goal* or *intention*, rather than the current state of affairs. It shares with the current body–world schema the feature that its contents are periodically updated in response to conscious acts of attention.

Each item in the body–world schema (current, projected, and historical) has a certain key part, which is part of the directive, or instruction, that led to the placement of that item in the schema. Thus if I consult my body–world schema to find out what I just saw on my right, the instruction by which I can reconfirm or update that item is immediately available through association: upon releasing an inhibition this instruction becomes carried out by the unconscious levels of processing.

When I choose to raise my arm I also generally choose, or intend, at the same time, to monitor its motion. Thus, just as for the current body–world schema, an item placed in the projected body–world schema can generally contain an *instruction* of the same kind as the instruction that produced that item. This instruction placed in the projected body–world schema will, if not amended, normally be carried out at the appropriate time by the unconscious processes.

It appears from these considerations that the brain can, under suitable conditions of alertness, sustain a "top-level process" with the following three general characteristics:

1   Its elements are events that actualize *instructions* to lower-level processes.
2   These instructions cause the lower-level processes to gather information, prepare for and execute actions, *and construct the next top-level instruction.*
3   Each top-level instruction is an updating of the body–world schema, or of some generalization of that schema.

At the neural level this sort of arrangement can be implemented by a category of patterns of neural excitations that I call "symbols". Each top-level instruction consists of a collection , or "chord", of these symbols. Each such symbol when "released" (e.g., by blocking some inhibitory signals) tends to activate, by association, the lower-level processes that it symbolizes.

This general picture of brain operation, which will be amplified later, appears compatible with the growing body of evidence coming from the brain sciences (see reference 27). I shall not review the evidence here but shall simply accept this overall picture and proceed to explore the impact of treating quantum mechanically certain important atomic processes that occur in the alert brain.

5.4 Incorporation of Quantum Mechanics

An element of brain dynamics where atomic processes play a key role is the release of the contents of a vesicle containing neurotransmitter into a synaptic junction. Our theoretical picture[5] of this process is that an action-potential pulse opens channels for calcium ions, which then migrate by diffusion to release sites. Several such ions must attach at a site to effect the release.

In the model of reference 5 a calcium ion travels about 50 nm in a time of about 200 μs, on its way from channel exit to release site. Simple estimates of the uncertainty principle limitations upon body-temperature calcium ions diffusing in this way show that the wave packet of the calcium ion must grow to a size many orders larger than the size of the calcium ion itself. Hence that the idea of a single classical trajectory becomes inappropriate: quantum concepts must in principle be used.

According to quantum theory the quantum state generated by this process of diffusion is a complex multiparticle state whose one-particle probabilities should approximate the probabilities given by the classical calculation.

The probability for an action-potential pulse to release a vesicle at a cortical synapse appears to be about 50%.[6] If, in some small time window (say a fraction of a millisecond), N synapses receive action-potential pulses then there will be 2^N *alternative* possible configurations of vesicle releases, each with a roughly equal probability. *Each alternative possibility is represented in the evolving quantum-mechanical wave function.*

The brain is a highly nonlinear system with feedback. Classical computer simulations show that the macroscopic state into which it will evolve is very sensitive to small variations at the synaptic level.[7] It is therefore, I think, virtually inconceivable that a variation over the 2^N alternative possible configurations of vesicle releases could, in general, have no influence on the eventual macroscopic state into which the system evolves. Nonlinear systems are generally very sensitive to small changes, and there is no reason to believe that the brain could be totally insensitive to such differences. Thus a universe containing a conscious brain, represented quantum mechanically, must be expected to evolve into a state that represents a *superposition of macroscopically different alternative possibilities for the brain*, provided there is no actual event that reduces the state to one that is not a superposition of this kind.

This raises the key question: At what point does an actual event intervene?

This issue was addressed by John von Neumann in an analysis that constitutes the foundation of the quantum theory of measurement.[8] Von Neumann considered a sequence of measuring devices, with the first one measuring the atomic system, and each other device measuring the response of the one before it, with the last one being, conceptually at least, some innermost level of the brain. He showed that, for his idealized case, it made hardly any difference at all at which point the actual event intervened to select one of the several macroscopically different possibilities: the quantum-mechanical probabilities were virtually independent of where the "Heisenberg cut" was drawn between the "quantum system" and the classically described device that was measuring it.

I shall exploit von Neumann's result by assuming that in the alert brain the main actual events occur at the point where a choice is made between *alternative possible instructions in the top-level process*. Since top-level instructions generally initiate large and differing responses by the lower-level processing mechanisms, this assumption is analogous to Heisenberg's assumption for inanimate objects that the actual event occurs only at a high level, where it chooses between states corresponding to *macroscopically different* actions of the object, such as the firing or nonfiring of a Geiger counter. Human conscious events are assumed to be the feels of these top-level events, which actualize *macroscopic* patterns of neural activity. We now have in place a general description of brain operation compatible with quantum theory, and can pose the question of the connection of brain to consciousness.

5.5 Brain and Consciousness

We are not conscious of what is going on in our brains. We are conscious of, for example, Beethoven symphonies, and sunsets. How can such a felt experience be the "feel" of some events in the brain?

To start with something simpler than a Beethoven symphony consider a triangle: Why, when we look at a triangle, do we experience three lines joined at three points, and not some pattern of neuron firings?

To answer this question let us consider first Edelman's explanation of how the visual cortex comes to be organized. The problem is this: the growth of the neurons connecting the retina to the visual cortex is not completely determined by genetic programming: there is a great deal of contingency. But then how does the structural information present at the retina get properly reconstituted at the cortex, rather than becoming hopelessly scrambled by the randomness of the neural connections?

The answer is that the saccadic movements of the eye cause the neurons that receive signals from adjacent retinal regions to receive *temporally* correlated signals. The resulting spatially distributed but temporally correlated patterns of excitation in the visual cortex then become automatically associated, by the facilitation process. Thus some of the structure at the retinal level becomes mapped into a spatially distributed *analog* structure in the realm of the cortical patterns of excitation.

Building up from this initial organization, initiated by the saccadic eye movements, repetitious patterns occurring at the retina *facilitate* corresponding patterns in the cortex.

Thus even though the neural wiring is somewhat haphazard, the process of facilitation nevertheless automatically establishes *analogs* of attended or recurring retinal patterns within the realm of the cortical patterns of excitations.

Patterns present in the visual cortex become associated, in the same way, with the neural accompaniments of those motor actions that bring them into being. Thus recurring features of the external visual scene will come to be associated with complex patterns of excitations that include the patterns that produce the motor actions that allow these features to be sensed.

Owing to this mapping of structure the cortical patterns generated by attention to the external triangle will be "congruent" to the external triangle. For example, the adjacency properties of the points along the three lines of the triangle will have their symbolic representations among the cortical patterns originally facilitated by the saccadic eye movements. Similarly, the various other perceived structural features of the external triangle will be represented by symbols that have been previously constructed by brain processes to represent such connections.

The act of attending to the external triangle implants this symbolic representation of the external triangle into the body–world schema. More specifically, this act of attending leads to an actual event that updates the body–world schema by actualizing an integrated chord of symbols that is "congruent" to the external triangle, in the sense that it will contain symbols that are the analogs of the various structural features that characterize the external triangle itself.

It might seem that this shift from the external triangle to a congruent inner representation has not helped at all, but only made things worse. Even if we grant the congruency property, the question remains: Why do we experience the *triangle* rather than the firings of neurons? We do not wish to introduce a homunculus that surveys the brain, and is able to decipher its complex activity and see a triangle.

This deciphering problem arises, however, only if one slides back to the classical concepts. In the quantum ontology a brain attending to an external triangle is not performing the retrograde act of transforming an actual external triangle into some congruent structure of particle motions, which must then be deciphered to be perceived as a triangle. Rather it is transforming the external triangle, which exists only as a pattern of *disjoint* events and tendencies, into a *single event* that actualizes, in integrated form, an image of the structural connections that inhere in the perceived triangle. The brain, therefore, does not convert an actual whole triangle into some jumbled set of particle motions; rather it converts a concatenation of separate external events into the actualization of some single integrated pattern of neural activity that is congruent to the perceived whole triangle. The central question is then: Why is the actualizing of this integrated pattern of activity felt as the perceiving of the triangle? More generally: Why do brain events feel the way they do?

5.6 Qualia: The Experiential or Felt Quality of Actual Events

The present theory asserts that each human conscious experience is the feel of an event in the top-level process occurring in a human brain. This brain process is asserted to consist of a sequence of Heisenberg actual events called the top-level events. Each such event actualizes some macroscopic quasi-stable pattern of neural activity. The pattern actualized by a top-level event is called a symbol. It normally consists of a set of other symbols, called its *components*, linked together by a superposed neural activity.

Actualizing a symbol S engenders enduring physical changes in the synapses (facilitation) that cause any subsequent actualization of any component of S to create a pattern of dispositions for the activations of the other components of S (association). Thus the actualization of any symbol S creates a pattern of dispositions for the activation of all symbols having a component that is a component also of S.

The actualization of any symbol S thus produces tendencies for the activation of various collections of symbols. One such collection, C, may be far more strongly disposed to activation than the others. Then the actualization of S constitutes an instruction for the actualization of that collection of symbols C.

Owing to quantum indeterminacy many alternative possible collections C must have nonzero weight. The next top-level event actualizes one collection, together with a superposed structure of neural activity that grows up around it and gives the whole pattern stability and distinctiveness, allowing it to stand out from the chaotic continuum of background activity and be actualized as a distinct quasi-stable pattern of neural activity. The full set of symbols, and of dispositions of symbols to activate symbols, created during the life of the brain by the top-level process, is called the *generalized body–world schema*. The body–world schema mentioned earlier is an integral part of it. Each top-level event augments the generalized body–world schema, and is therefore an updating of it.

The generalized body–world schema is an organizational structure in which all symbols are effectively stored, in latent form, for later retrieval by cross-referencing. The retrieval mechanism is presumably this: If a symbol S has a disposition to be activated by several symbols, then the simultaneous actualization of these several symbols will cause S to be activated more quickly, and hence become actualized *before* the symbols less strongly disposed to activation reach the threshold for possible actualization.

This retrieval mechanism can allow brain process to actualize, by cross-referencing, the symbol that represents, for example, the occupant of a certain place at a certain time, without interference from the symbols representing the occupants of that place at other times, or the occupants of other places at that time; and to actualize the symbol that represents the place where an object represented by a certain symbol is located at a certain time, without interference from the symbols representing the locations of that object at other times. The generalized body–world schema thus becomes the physical basis for the long-term memory system. The top-level process is the generator of this memory system.

We may now state an essential point: Each top-level event actualizes a symbol, and this symbol has *components* that are themselves symbols. Thus each top-level event is represented by a symbol that has a *compositional structure*: it has components that are entities of the same kind as itself.

Consider next the mental side. The structure of mental states has been extensively studied. I accept the conclusions of William James, who cites with strong approbation the following quotation:

> Our mental states always have an *essential unity*, such that each state of apprehension, however variously compounded, is a single whole of which every component is, therefore, strictly apprehended (so far as it is apprehended) as a part. Such is the elementary bases from which all our intellectual operations commence.[9]

A component of a thought, so far as it is apprehended, is itself a possible thought. Thus each thought has a compositional structure: it has components that are entities of the same kind as itself. Our basic principle is that the compositional structure of the feel of a top-level event is isomorphic to the compositional structure of the symbol actualized by that event: there is a one-to-one mapping of symbols to feels, and this mapping preserves compositional structure.

The fundamentally integrative character of the Heisenberg actual event enters here in a critical way. The Heisenberg event *grasps as a whole* an entire integrated pattern of physical activity. This essential unity of the actualized physical state accords with the essential unity of its mental counterpart.

William James has described the profound conceptual inadequacy of classical mechanics—as a basis for understanding the connection between brain and mind—that is so satisfactorily resolved at this point by quantum theory. Having emphasized the essential unity of each thought, and a first difficulty that arises from it, James goes on to say:

> The second difficulty is deeper still. *The "entire brain-process" is not a physical fact at all.* It is the appearance to an onlooking mind of a multitude of physical facts. "Entire brain" is nothing but our name for the way in which a million of molecules arranged in certain positions may affect our senses. On the principles of the corpuscular or mechanical philosophy, the only realities are the separate molecules, or at most the cells. Their aggregation into a "brain" is a fiction of popular speech. Such a fiction cannot serve as the objectively real counterpart to any psychic state whatever. Only a genuine physical fact can so serve. But the molecular fact is the only genuine physical fact...[10]

In the quantum ontology the only genuine physical facts are the actual events. Hence some actual event must "serve as the objectively real counterpart to [each] psychic state". But in this case the essential unity of the psychic state—so incomprehensible within reductionist classical thought—mirrors the essential unity of its physical counterpart. In both cases the ontological progression is from the ontologically fundamental wholes to their ontologically subordinate components, rather than from presumed ontologically fundamental elements to assemblies thereof. This shift from synthetic ontology to analytic ontology is the foundation of the present work.

A fundamental feature of experience is the feel of the "flow of consciousness", or the "perception of time". On the other hand, each actual event is ontologically distinct from all others, and its feel is the feel of itself alone. Thus the "present" mental event is the feel exclusively of the "present" physical event; it has no access to past physical events.

But how, then, does one account for the "flow of consciousness" and the "perception of time"? These phrases refer to an extensively analyzed empirical structure described in rough terms by William James in the following way:

> If the present thought is of ABCDEFG, the next one will be of BCDEFGH, and the one after that of CDEFGHI—the lingerings of the past dropping successively away, and the incomings of the future making up the loss.[11]

According to this picture, each *immediately present* mental event contains *within itself* a sequence of parts perceived as "temporally" ordered.

This "temporal" structure of each mental event evidently arises, in part, in the following way: owing to the quasi-stable character of symbols the symbol actualized by a top-level event will generally have among its components, many of the components of the symbol actualized by the preceding top-level event: the set of components of the new symbol will include many of the components of its predecessor, together with some new symbols. Thus the feel of the new event will have components that correspond to components of earlier events.

If someone recites quickly an unfamiliar sequence of four numbers, an attentive listener can readily repeat the sequence, or repeat the part of it starting from any one of its four components. However, reciting the sequence in reverse order requires more effort. Thus there is evidently a dynamical tendency for associations between the temporal slices of a thought to move from any slice to its temporal successor, rather than randomly about. The existence of this tendency means that the superposed structure of the symbol, which creates the dispositional "associations" between its components, must give larger dispositions to the associations that run forward in the "temporal" ordering. Since this enveloping neural structure tends to recreate the earlier temporally ordered patterns of activity, such a biasing for "forward" association is to be expected. It will be accommodated in our representation of the compositional structure of a symbol in terms of its components by allowing both a "$+$" composition that is commutative $[a + b = b + a]$ and also a "sequential product" that is noncommutative and nonassociative $[(abcd) \neq (abdc), ((abc)(def)) \neq (abcdef)]$. The "$+$" composition combines symbols without "temporal" biasing, and the sequential composition combines symbols with "temporal" biasing. Thus the brain event that follows upon the hearing of the spoken sequence $(5, 6, 2, 8)$ is represented by (5628), and it is felt as the heard sequence $(5, 6, 2, 8)$. Here I have used the same numeric symbols to represent the spoken words, the components of the symbol actualized by the top-level brain event, and the components of the feel of that event.

James's picture of a marching sequence of *fixed* letters is only a first approximation. Each actualized symbol creates dispositions for the activation of various symbols that were actualized together with itself in earlier top-level events. Thus as one of James's letters marches through the sequence of successive events its original symbolic counterpart becomes embellished by an expanding network of symbols, consisting of symbols that were actualized together with it during earlier top-level events. The feel tied to the marching letter consequently becomes embellished by the feels of these earlier events: its "meaning" becomes enlarged and sharpened by the agglutination of feels associated with related past events.

The symbols are quasi-stable structures with fatigue characteristics that cause them eventually to fade out. Thus after an initial period of intensity, accompanied by a growing sense of "meaning", the feel tied to any fixed letter in James's picture will begin to fade out, and it will eventually die away. James has described this waxing and waning of the intensity and sharpness of the "temporal" components of a present mental state, when it is analysed in terms of the variation of the "temporal" variable: the newest components are still vague, the ones later in the "temporal" sequence are clearer, and the older ones fade away.

The range of possible "meanings", as characterized by the number of possible structural forms of these embellishments, can be huge. James cites evidence that a mental event may have as many as 40 temporally distinguished parts.[12] Suppose there are just ten fundamental symbols, and that all others are formed by simply the sequential compositions of these ten. Then the number of possible embellishments generated in the first 20 steps, e.g., before the fading sets in, is 10^{20}.

Embellishment leads to "meaning" because the embellished symbol is experienced as a felt structure of feelings each of which corresponds to a related past event: an observed "bicycle" comes to be associated with a structure of feels in which are imbedded childhood experiences of locomotion, spills, adventures, etc., i.e., of what a bicycle *led to* in the past, and hence might lead to again.

These "meanings" arise, however, only from the *structural content* of the symbol: the ten basic symbols act as undefined symbols from which all the structures are built, but the 10! permutations of these ten basic symbols leave the internal structural content unchanged: all *connections* between feels are left unchanged by these permutations. Thus the possible shades of meaning number, in principle, $10^{20}/10!$, in this example.

The distinction being emphasized here is between the elemental, or absolute, units of experience, such as the immediate direct experience of redness, or of the pitch of high C, and the meanings of symbols that arise from their compositional structures. The former are the feels of certain actualized patterns of neural activity, and would be different if the patterns of neural activity representing these symbols were different. The latter reside in the internal structural composition of the symbol and would be left unchanged if the feel of all symbols were shifted in a way that maintained the feel of "nearness" that feels can have to one another.

The dynamical process of embellishment considered above, in which the symbolic counterpart of each "letter" in the temporal sequence develops associations by itself, as if it developed in isolation from the other symbols actualized together with it, is an over simplification: symbols actualized together act together; they act as combined dispositions for the activation of other symbols. It is this capacity of the different temporal components of a single top-level event to act jointly that gives brain process its capacity to compare and combine symbols, and to manipulate them in other ways.

Each normal top-level event contains a background of symbols that persists through the various "temporal" slices into which it is divided. This background of symbols is felt as a persisting background of intentions and other feels, against which the more transitory feels are contrasted. This background constitutes the feel of "self" that pervades each normal human experience.

This felt "self" is simply *part* of the experience. The only *carrier* that links these experiences together is the brain: the brain is the only *receiver* of the experiences. Each experience exists, and has a structure that mirrors the structure actualized in the brain by the event it reifies. What could be more simple and natural?

5.7 Comparison with Other Treatments

Gerald Edelman and John Eccles have set forth detailed proposals concerning the connection between mind and brain. Their proposals, which constitute serious efforts to accommodate, and integrate, the growing body of neurophysiological, neuropsychological, and other relevant scientific data, are compared in this section to the theory described above. Comparison is made also to the positions of Bohr, von Neumann, and Wigner.

5.7.1 Comparison with Edelman

Edelman's theory rests on a comprehensive general account of the development of the brain during evolution, during embryonic growth, and during the life of the individual person. This account provides a fairly detailed description, based on the relevant scientific data, of the development and functioning of a level of neural processing that is subsumed in my term "lower-level processing".

According to Edelman's theory these lower-level brain processes must contain four specific components if consciousness is to emerge. The first of these components is *perceptual categorization*, which is a neural process mediated by synaptic change that causes particular patterns of neural activity to become activated by, and hence associated with, particular patterns of signals from sense organs. The second lower-level component deemed necessary for the emergence of consciousness is the functioning of neural pathways dedicated to the incorporation into brain processing of *values* pertaining to the physiological and other needs of the organism.[13] The third necessary lower-level component is *memory*, which, in this context, is a system property of the brain, mediated by synaptic change, which arises from the continual creation of new patterns of neural activity representing new categories. These new categories, expressed as neural activities, correlate and compare the categories previously created.[14] The fourth component of lower-level neural processing deemed necessary for consciousness is a component that affects *learning*, which is

> context-dependent behavioral change governed by positive or negative value under conditions of expectancy.[15]

These four components of brain processing can, according to Edelman's theory, function without the occurrence of conscious awareness, i.e., without consciousness. According to Edelman's theory,

> consciousness is the result of an ongoing categorical comparison of the workings of two kinds of nervous organization. This comparison is based on a special kind of memory, and is related to the satisfaction of certain physiologically determined needs as that memory is brought up to date by the perceptual categorizations that emerge from ongoing present experience. Through behavior and particularly through learning, the continual interaction of this kind of memory with present perception results in consciousness.[16]

The terms "memory" and "perception", as used here, do not in themselves carry any connotation of conscious awareness: they pertain to *neural* process, as described above. Consciousness is thus claimed to be the *result* of an interaction between these two components of the unconscious neural processing.

This key process of the emergence of consciousness is described in various places in Edelman's book:

> Imagine that the various memory repertoires dedicated to the storage of the categorization of *past* matches of value to perceptual category are [reciprocally] connected to [the neural systems] dealing with *current* sensory input and motor response. By such means, past correlations of category with value are now interactive in real time with current perceptual categorizations *before* they are altered by the value-dependent portions of the nervous system. A kind of bootstrapping occurs in which current value-free perceptual categorization interacts with value-dominated memory before further contributing to alteration of that memory. Primary consciousness thus emerges from a . . . recategorical memory (relating *previous* value-category sequences) as it interacts with current input categories arising from neural systems dedicated to present value-free perceptual categorizations.[17]

> It is the discriminative comparison between value-dominated memory involving the conceptual system and the current ongoing perceptual categorization that generates primary consciousness of objects and events . . .[18]

> . . . the generation of a "mental image" . . . emerges as a result of a series of . . . correlations of [perceptual] categories to . . . values . . .[19]

> The functioning of these key [reciprocal] connections [between past value-category connections and current perceptual categorizations] provides the sufficient condition for the appearance of primary consciousness.[20]

The question arises as to how one is to interpret this claim that this special neural process is a *sufficient condition* for consciousness to occur. Does this claim mean that the occurrence of consciousness is *logically entailed* by the occurrence of this neural process?

At the beginning of his book Edelman lists a set of constraints on his undertaking. The first of these is the condition that

> any adequate global theory of brain function must include a scientific theory of consciousness, but to be scientifically acceptable it must avoid the Cartesian dilemma. In other words, it must be uncompromisingly physical and be based on *res extensa*, and indeed be derivable from them.[21]

This condition seems to demand that the emergence of consciousness be *derivable* from the properties of matter. Edelman accepts

> modern physics as an adequate description for our purposes of the nature of material properties.[22]

Thus Edelman's demand appears to be that the emergence of consciousness must be actually derivable from physics, or at least from properties of systems describable in principle in terms of the concepts of physics. This strong interpretation is reinforced by the claim made in the final chapter that no special addition to physics is required for the emergence of consciousness.[23]

If this indeed be the claim, then Edelman's account falls short. For the particular neural process that is claimed to be sufficient for the emergence of consciousness is a physical process describable in principle in terms of neural patterns of excitation, and hence, if one ignores the subtleties connected with quantum theory, as Edelman does, in terms of atoms, and electrons, etc. According to the precepts of physics (if quantum effects are ignored) these atoms and electrons, etc., will behave in exactly the same way whether or not a quality of conscious awareness emerges in connection with this particular physical process.

This particular neural process may be connected in some very natural way to some particular quality or kind of awareness. However, that fact, joined to the laws of physics, does not *entail* that this particular quality of awareness must actually come into existence when that physical process occurs. Consequently, the assertion that this quality of awareness does come into existence under those special physical conditions is "a special-addition physics": it is not entailed by, or derivable from, the principles of physics.

To the extent that one ignores the effects introduced by quantum theory, and hence adheres to the precepts of classical physics, this extra or added quality of awareness is necessarily nonefficacious: it has no effect on the ongoing neural process. The theory therefore does not succeed in avoiding the Cartesian dilemma, as the initial condition demanded, but introduces a causally disconnected *res cogitans*.

Edelman has, it appears to me, accepted a tacit assumption that if there is a neural action that *functions* in a way that is a natural image of the subjective *feel* of a possible conscious event, then this conscious event will in fact occur if the neural action occurs. This is Edelman's implicit analog of my explicit postulates about feels.

The problem with Edelman's approach is that if one adheres to his demand that

[the] view of brain function and consciousness should be based on materialist metaphysics,[24]

and hence rules out quantum physics, and perforce retreats to classical physics, then there is nothing in the *physics* that singles out these special processes as being in any way special. They are special only because they can be associated in a certain way with things outside classical physics, namely possible conscious experiences. But then the claimed connection between these two domains is, from the physics point of view, completely ad hoc. This ad-hocness is connected with the fact that the conscious awareness, per se, is, within the conceptual framework of classical physics, wholly nonefficacious.

In the Heisenberg quantum ontology, on the other hand, the place where consciousness enters is, from the *physics* point of view, *dynamically singled out*, and consciousness is able to become causally efficacious. Consequently, the quantum theory of consciousness comes much closer to filling Edelman's demand that the theory be based on *res extensa*, as described by modern physics, than his theory does.

5.7.2 Comparison with Eccles

The theory of Eccles[25] is explicitly dualistic: it postulates a mental entity that *interacts* with the brain, and that continues to exist after the death and destruction of the brain. This "homunculus" is allowed to influence brain process by exploiting the lack of determinism allowed by quantum theory. Although Eccles's theory thus exploits the freedom introduced by quantum theory, it neither appeals to, nor exploits, the profound conceptual change wrought by quantum theory.

Eccles's theory is fundamentally different from the theory proposed here, which explicitly ties *every* human conscious event to a corresponding physical event in a human brain. Neuropsychological evidence exists that discriminates, I believe, between Eccles's theory and mine. It comes from the behavior of certain patients who have suffered massive parietal-lobe damage, and subsequently exhibit a neglect syndrome: a loss of ability to attend to certain parts of their bodies located contralateral to the damaged area of the brain. Their behavior suggests that the impairment is more than just a loss of ability to control or sense parts of the body, or even to communicate or speak about them, but is rather a complete disappearance of any representation of the afflicted part of the body from the patient's repertoire of conscious thoughts: the afflicted part seems simply to disappear from the patient's conception of his body.

Such an effect can be naturally understood as a consequence of elimination of the representation of the afflicted part from the body schema by the destruction of the neural basis of the patterns of activity that constitute the symbols that correspond to that part of the body. In the quantum theory of consciousness proposed here the mental universe of each human being consists exclusively of the felt quality of actual events constructed out of the symbols that are the building blocks of the (generalized) body–world schema: consciousness is the felt quality of the manipulating actions of these symbols upon each other. These symbols are thus the *currency of consciousness* and the destruction of any of them must cause a reduction in the person's mental universe.

A homunculus residing in a separate mental world, and able to survive the death and destruction of the brain, would, presumably, not be itself impaired by the brain damage: *its* mental universe would be left essentially intact. The damaged brain would be unable to respond as fully to the action of the homunculus upon it, and this impairment would result in problems in communication, and control, and in the reciprocal action of sensing. But the representation of the afflicted part would not disappear from the patient's mental universe itself, as is suggested by the evidence: the patient should not be *puzzled* to discover that there is a left arm connected to his body;[26] the patient should "know" that he has a left arm, even though he has recently been deprived by brain damage of the ability to directly sense or control it. Hence he should not be *puzzled* to discover it.

Some other evidence supportive of the quantum theory, but not necessarily discriminative relative to Eccles's theory, is the data of Libet[27] pertaining to the *delay* in the occurrence of the conscious awareness of a voluntary intention to act, relative to the onset of the neural activity that prepares for the conscious event. The foundations of the quantum theory of consciousness are: (1) the idea that the brain functions to plan, select, and execute single integrated actions; (2) the idea that, owing to the unavoidable intrusion of quantum uncertainties into the synaptic processing, and the subsequent amplification of these quantum synaptic processes, the brain functions in a way that is basically similar to a quantum measuring device such as a Geiger counter, in the specific sense that the evolution of the physical system in accordance with the basic local law of evolution (i.e., the Schrödinger or Heisenberg equations of motion) necessarily produces, normally, a state that represents a *superposition* of macroscopically distinctive states, such as the firing or nonfiring of the Geiger counter, or the activation or nonactivation of the neural activities that represent the intention to raise an arm; and (3) the acceptance of Heisenberg's position that these two alternative macroscopic possibilities *do not both actually occur*, in some absolute sense, as is claimed by the competing "many-worlds" interpretation of quantum theory, but that, instead, the representation of the physical system by a quantum-mechanical state is a representation not of the *actual* world itself, but rather of the *tendencies* for the occurrence of an actual event that will select and actualize *one* of the macroscopically distinct alternatives.

In the context of the Libet experiments the critical point is that according to the Heisenberg picture there must *first* be a separation, generated by the evolution in accordance with the deterministic equation of motion, of the physical state into parts representing several macroscopically distinct possibilities *before* the act of choosing one of these macroscopically distinct alternatives occurs. In the brain most of the processing activity is done at the unconscious level: the lower-level process first prepares the distinctive alternatives, and the Heisenberg actual event then selects and actualizes one of them. Thus the delay found by Libet is demanded by this quantum-mechanical theory of consciousness.

In the homuncular theory it would seem that the homunculus could *first* decide to raise the arm, and then interact with the brain in order to bring about its desired end, and that the conscious event would therefore *precede* the neural activity that leads to the motor action.

5.7.3 Comparison with Bohr

The strictly orthodox Copenhagen interpretation of quantum theory brings human experience into physics in a way much more explicit than classical theory did. The quantum theory is interpreted as *fundamentally* a theory that allows the scientist to form expectations about certain of his experiences. These are experiences that can be described in terms of specifications formulated in terms of the concepts of classical physics. This last stipulation effectively removes the individual human experience from any place of prominence, for it makes the referents of the theory a class of external facts that all observers generally agree upon. So, in the end, the role of the subjective observer is no different than it was in classical physics: he is the subjective observer of essentially objective external facts. The issue of the connection of brain processes to mental process is thus never brought into question. In fact, this issue is moved by Bohr outside the domain to which quantum theory might apply by raising certain objections in principle to the application of quantum theory to biological systems.

In Bohr's words:

> The incessant exchange of matter which is inseparably connected with life will even imply the impossibility of regarding an organism as a well-defined system of material particles like the system considered in any account of the ordinary and physical chemical properties of matter. In fact, we are led to conceive the proper biological regularities as representing laws of nature complementary to the account of properties of inanimate bodies . . .[28]

The problem behind these words is that the interaction of a quantum system with its environment introduces conceptual difficulties that are, in fact, much more severe than those of classical physics. In classical physics when a particle leaves the system and becomes part of the environment it leaves the system in a state that is well defined in principle. In quantum theory this is not the case. The state of the residual system alone is not well defined: one must, in principle, for a complete description, keep track of each particle that has left; the state of the residual part depends on the location of each particle that has left, but each such location is defined only as a smeared-out superposition of possibilities. This means that each current brain state is not a single state in which the parts have well-defined locations, but is rather a superposition of states in which the parts have locations that depend on the ill-defined locations of the many particles that have long since left the brain and body. But what thought can be associated with such a smeared-out superposition of brain states?

As Bohr emphasizes, some new ideas are needed: the strictly orthodox interpretation of quantum theory gives neither a practically useful nor conceptually cogent picture of what is going on in brains. The Heisenberg ontology provides the simplest cogent extention of the strictly orthodox position. In it the actual brain events constitute a closely packed sequence of events that continually redefine the key macroscopic features of the brain state.

5.7.4 Comparison with von Neumann and Wigner

Von Neumann's analysis of the process of measurement involves a sequence of measuring devices, each of which detects the result of a measurement performed by the device prior to it in this sequence, with the final "device" lying deep within the brain. Von Neumann accepted a principle of "psychophysical parallelism", which asserts that the process of subjective perception has a counterpart in the objective physical world, described in ordinary space.

Von Neumann's colleague, Eugene Wigner, elaborated upon this idea, suggesting, rather, a reciprocal *interaction* between mind and matter.[29] However, in his later works[30] Wigner rejected the idea that unmodified orthodox quantum theory can be applied to *macroscopic* systems. He, like Bohr, cited the important effects of interactions with an uncontrollable environment.

It is worth emphasizing that in the proposal being advanced here the actual events associated with human conscious experiences are not presumed to be the *only* actual events: actual events associated for example with the firing of a Geiger counter are presumed to exist, as Heisenberg assumed. Here it is merely accepted that, under similar conditions, the brain, which *also* is a physical system, should *also* be subject to the collapsing action of actual events.

5.8 Related Philosophical Issues

The success of classical physics in earlier centuries gave credence to the Newtonian idea of the universe as a machine, and to the concomitant Cartesian idea of consciousness as an impotent witness to a *preordained* course of events. The rise of quantum theory in this century modified the Cartesian idea only slightly. In the absence of a quantum-mechanical treatment of the brain, consciousness became, instead, an impotent witness to a *whimsical* course of events. This constitutes no basic change in the Cartesian conception of the role of consciousness.

This Cartesian idea, backed by the authority of science, has exerted an enormous influence on philosophy, and a corrosive influence on the philosophical foundations of human values. On the other hand, the quantum theory of consciousness described above will, if validated by ongoing empirical studies, constitute a scientifically supported alternative to the Cartesian ontology. It will, as such, have far-reaching philosophical ramifications. Two of these are briefly mentioned.

5.8.1 The Efficacy of Consciousness

In Heisenberg's ontology the actual event is efficacious: it actualizes one localized macroscopic pattern of activity from among a set of previously allowed possibilities. These possibilities, or, more precisely, the tendencies for the actualization of these alternative possible activities, are generated in a mathematically deterministic way by Heisenberg's equations of motion, which are the quantum analogs of corresponding classical equations of motion.

According to the theory advanced here each actual event has two aspects; a feel, and a physical representation within the quantum formalism. The feel is asserted to be a veridical image of the effect of the action of the physically described event.

At the purely physical level the Heisenberg actual event is passive: it is simply the coming into being of a new set of tendencies. However, in the context of the present ontology the actual event must be construed actively: the event *actualizes* the shift in tendencies. If the feel is identified as the active aspect of the event, then the feel is the veridical feel of actively actualizing the new state of affairs, and consciousness becomes the efficacious agent that it veridically feels itself to be.

5.8.2 The Quantum Choice

The question arises: What determines *which* of the alternative possible brain activities is actualized by an actual event?

According to contemporary quantum theory, two factors contribute to this quantum choice. The first is the local deterministic evolution of tendencies governed by the Heisenberg equation of motion. This factor brings in all of the local historical influences such as heredity, learning, reflective contemplation on priorities and values, etc., that contribute to the formation of the current state of the brain. These factors determine, however, only the *tendencies*, or *weights*, associated with the various possible distinct courses of action. Then an actual event occurs. This event actualizes one of the distinct top-level patterns of brain activity, and hence selects one of these distinct possible courses of action. This selection is, according to contemporary quantum theory, made by the second factor: pure chance.

Pure brute stochasticity, with no ontological substrate, is in my opinion an absurdity: the statistical regularities must have some basis. On the other hand, the answer provided by contemporary quantum theory is probably correct in the sense that the basis for the quantum choices cannot be conceptualized in terms of the ideas that it employs. Within that framework these choices *must* therefore appear to come out of nowhere; they must be, in the word used by Pauli and by Bohr, "irrational".

This inadequacy of the usual concepts can, I believe, be deduced by attending to certain features of the mathematical structure of the quantum formalism itself. The Heisenberg ontology is a kind of pictorial representation of this mathematical structure. It has, however, one exceedingly strange feature. This feature is superficially similar to the correlation effects that occur in classical statistical mechanics. Classically, if two systems become statistically correlated, owing to some interaction between them, and each of them subsequently moves to one of two regions that are spatially well separated, then a measurement on one of the systems can provide statistical information about the other system, even though the two systems are far apart. There is nothing strange about this. However, if the statistical weights are interpreted as "objective tendencies", which have *objective* existence, which is the basic idea of Heisenberg's ontology, then the change in the far-away statistical properties as a consequence of a measurement performed here would constitute an instantaneous action-at-a-distance.

The Heisenberg ontology manifests precisely such an action-at-a-distance, and hence would seem to be unacceptable. At least it seemed to be unacceptable until the work of J. S. Bell in 1964.[31] That work, suitably reformulated,[32] shows, however, that if the choices between macroscopically distinct alternatives, such as the firing or nonfiring of a Geiger counter, are indeed made by nature, as the Heisenberg ontology maintains (in opposition to the many-worlds view, which maintains that *both* alternatives occur, but in noncommunicating branches of the universe), then these choices *cannot* be implemented by local actions: they can be implemented *only* by actions that transcend spacetime separation, i.e., that can act without attenuation over large spacelike distances.

The conclusion, here, is that if the many-worlds idea is incorrect, and the *macroscopic* world is therefore roughly what it appears to be, then the structure of the *predictions* of quantum theory itself demands that the basic process of nature be intrinsically global: it *cannot* respect spatial separations in the way that familiar causal processes do. Thus to the extent that we confine our thinking to processes of the familiar local kind the quantum choice *must* appear to come from nowhere.

The implication of the foregoing considerations is that although the flow of conscious events associated with a particular human brain has important personal aspects, which arises from the fact that the content of these events is the feel of the acts of manipulation of the web of symbols created by the brain upon that web itself, nevertheless the fundamental process that is expressing itself through these local events is intrinsically global in character: it cannot be understood as being localized in the brain, or in the body. Rather it must act in a coordinated way over much of space. Neither contemporary science nor the present work addresses the issue of how that global process works. Our ignorance concerning this intrinsically global process is represented in these theories by the introduction of "pure choice".

5.9 Summary

The quantum theory of consciousness developed here:

1   makes consciousness efficacious;
2   rests directly on the mathematical formalism of quantum theory;
3   parsimoniously accepts no kinds of entities not present in the Heisenberg or Copenhagen conceptions of nature;
4   adheres fully to quantum thinking;
5   meets the Einstein demand that basic physical theory describe the processes of nature, not merely our knowledge of those processes;
6   mends the Cartesian cut by identifying an entity, the Heisenberg actual event, that unites as its two faces the subjective and objective aspects of mind–brain action;

7   enunciates a principle of mind–brain isomorphism that seems able to account for the full content and structure of felt human experience, and its connection to brain process;

8   identifies the "self" as a slowly evolving background component of human experience, not as the owner of that experience;

9   describes the consciousness of man as a localized aspect of a global integrative process.

References

1 W. James, *The Principles of Psychology* (Dover, New York, 1950; reprint of 1890 text), vol. I, p. 146.

2 W. James, *Psychology: Briefer Course* (Henry Holt, New York, 1893), p. 468.

3 W. Heisenberg, *Physics and Philosophy* (Harper and Row, New York, 1958), chap. III.

4 G. Edelman, *The Remembered Present: A Biological Theory of Consciousness* (Basic Books, New York, 1990); *Neural Darwinism: The Theory of Neuronal Group Selection* (Basic Books, New York, 1987).

5 A. Fogelson and R. Zucker, Presynaptic Calcium Diffusion from Various Arrays of Single Channels: Implications for Transmitter Release and Synaptic Facilitation, *Biophys. J.* **48**, 1003–1017, (1985).

6 H. Korn and D. Faber, Regulation and Significance of Probabilistic Release Mechanisms at Central Synapses, in *Synaptic Function*, edited by G. Edelman, W. E. Gall, and W. M. Cowan (Wiley, New York, 1987).

7 L. Ingber, Statistical Mechanics of Neocortical Interactions: Dynamics of Synaptic Modification, *Phys. Rev. A* **8**, 385–416, (1983).

8 J. von Neumann, *Mathematical Foundations of Quantum Mechanics* (Princeton University Press, 1955), chap. VI.

9 Ref. 1, p. 241.

10 Ref. 1, p. 178.

11 Ref. 1, p. 606.

12 Ref. 1, p. 612.

13 Ref. 4, pp. 93–94.

14 Ref. 4, pp. 56, 109–112.

15 Ref. 4, pp. 93, 57.

16 Ref. 4, p. 93.

17 Ref. 4, p. 97.

18 Ref. 4, p. 155.

19 Ref. 4, p. 101.

20 Ref. 4, p. 100.

21 Ref. 4, p. 10.

22 Ref. 4, pp. 19, 253.

23 Ref. 4, pp. 19, 260.

24 Ref. 4, p. 10.

25 J. C. Eccles, *Proc. Roy. Soc. Lond. Ser. B* **227**, 411–428 (1986); A Unitary Hypothesis of Mind–Brain Interaction in the Cerebral Cortex: Dendrons, Psychons (the 5th Dennis Gabor Lecture); The Microsite Hypothesis of the Mind–Brain Problem (unpublished).

26 B. Williams, *Brain Damage, Behavior and the Mind* (Wiley, New York, 1976); D. L. Schacter, M. P. McAndrews, and M. Moscovich, Access to Consciousness: Dissociations between Implicit and Explicit Knowledge in Neurophysiological Syndromes, in *Thought Without Language*, edited by L. Weiskrantz (Clarendon, Oxford, 1988); K. Pribram, *Brain and Perception: Holonomy and Structure in Figural Processing* (John M. MacEachran Lecture Series, I. Erlbaum and Associates, Hillsdale, NJ, 1990). Professor Pribram, commenting on my paper, says "There is good evidence that the brain processes responsible for updating are co-extensive with those involved in processing conscious awareness of the *projected body–world schema*. This reflective aspect of 'self-consciousness' is called intentionality by philosophers [and] is dependent upon paying attention . . .". My theory claims, essentially, that *all* conscious awareness is a generalization of this kind of consciousness. Professor Pribram also suggests that my term "symbol" be replaced by the term "action pattern", to distinguish it from the AI use of the term "symbol". He also affirms that there is a wealth of neurophysiological and neuropsychological data to support my picture of brain organization in terms of a top-level process that issues instructions to lower-level processes as described in the text.

27 B. Libet, Cerebral "Time-on" Theory of Conscious and Unconscious Mental Function (unpublished).

28 N. Bohr, *Atomic Physics and Human Knowledge* (Wiley, New York, 1959), pp. 20–21.

29 E. Wigner, in *The Scientist Speculates*, edited by I. J. Good (Windmill Press, Kingswood, Surrey, England, 1967).

30 E. Wigner, in *Quantum Optics, Experimental Gravitation, and Measurement Theory*, *NATO ASI Series, Series B: Physics*, vol. 94.

31 J. S. Bell, *Physics* **1**, 195 (1964).

32 H. P. Stapp, EPR and Bell's Theorem: A Critical Review, *Found. Phys.* **21**, 1 (1991).

Part III

Implications

6 Mind, Matter, and Pauli

6.1 Introduction

Wolfgang Pauli was called by Einstein his "spiritual heir", and his unrelenting demand for precision and clarity earned him the title of "the conscience of physics". A godson of the great philosopher of science Ernst Mach, he was philosophically astute and, with Niels Bohr and Werner Heisenberg, a principal architect of the orthodox Copenhagen interpretation of quantum theory. This approach to the theory allowed physicists to avoid assigning paradoxical properties to nature. It did so by adopting a philosophically radical stance: regard atomic theory not as a description of atomic processes themselves, but rather as a description of connections between human observations. This renunciation of the traditional scientific ideal of erecting a coherent idea of physical reality was the chief objection against the Copenhagen view raised by Einstein. Though Einstein admitted that it was still unexplained why science had succeeded even as far as it had in creating a mathematical understanding of nature, he held that we must nonetheless persist in the endeavor: otherwise even the possible would not be achieved.

In a 1948 letter to his friend Marcus Fierz, Pauli writes:

> When he speaks of "reality" the layman usually means something well-known, whereas I think that the important and extremely difficult task of our time is to build up a fresh idea of reality.[1]

This idea was meant to encompass not only the material side of nature, but also its psychic or spiritual side:

> It seems to me—however it is thought, whether we speak of "the participation of things in ideas" or of "inherently real things"—that we must postulate a cosmic order of nature beyond our control to which *both* the outward material objects *and* the inward images are subject... *The ordering and regulating must be placed beyond the difference between "physical" and "psychical"* —as Plato's "Ideas" have something of the concepts and also something of the "natural forces...".

In a later letter (13 October 1951) Pauli goes on to say, in regard to the significance of the entry of a basic element of chance into physics:

Something that previously appeared closed has remained open here, and I hope that *new concepts* will penetrate through this gap in the place of [psychophysical] parallelism, and that they should be uniformly *both* physical *and* psychological. May more fortunate offspring achieve this.[2]

These ideas of Pauli appear to represent a fascinating reversal of his earlier position; the quantum element of chance is viewed no longer as a *veil* that must obscure forever our complete understanding of reality, but rather as an *opening* to a still deeper understanding. Yet Pauli's view is no mere conversion to the Einsteinian view that science should strive to represent physical reality. Einstein accepted the traditional scientific separation of mind from matter, whereas Pauli is suggesting that the element of chance in quantum theory provides an opening not to a traditional physical reality but rather to a reality lying beyond the mind–matter distinction.

My intention here is to explore this idea, which, if correct, would open up a whole new chapter in science. But before venturing beyond the confines of mind and matter it will be useful to review briefly the role of mind in modern science.

6.2 Mind in Classical Physics

The conceptual separation of mind from matter initiated by Descartes—and completed by other philosophers and physicists—rendered classical physics both reductionistic and local. It was reductionistic because the full description of the material world was reduced to a *collection of numbers*, and it was local because these numbers described *local* properties, such as (1) where particles are positioned at various times, and (2) the strengths of the electric, magnetic, and gravitational fields at various points in space and time. This local-reductionistic aspect allowed the material world, *as it was conceived of in classical physics*, to be brought, at least in principle, under full mathematical control. The thoughtlike aspects of nature, which Descartes placed in a separate realm called *res cogitans*, were eventually detached from the material world in a way that rendered them irrelevant to the course of material events.

Science, as conceived by Newton, was not a closed book: it was expected to grow and develop. Consequently, a scientific theory did not need to be *complete* in order to be useful and acceptable. For example, Newton did not explain the cause of gravity. He admitted to having tried and failed to find such a cause, yet affirmed his conviction that such a cause must nonetheless exist.[3] He left the unsolved problem to the consideration of his readers. Two centuries passed before a reader, Albert Einstein, found a satisfactory solution. Similarly, the omission of human experience from Newton's laws does not mean that the relationship of thought to matter must remain forever beyond the reach of science.

6.3 Heisenberg's "Ontologicalization" of the Copenhagen Interpretation

The Copenhagen interpretation of quantum theory is fundamentally episte-mological: it is concerned with our knowledge. The quantum-mechanical formulas

> merely offer rules of calculation for the deduction of expectations about observations obtained under well-defined experimental conditions specified by classical physical concepts.[4]

Within this mathematical framework, devised by quantum physicists, the central object is the so-called "wave function". It is considered to represent *our knowledge* of the physical situation, and it therefore changes suddenly when we receive new information and our knowledge therefore suddenly changes.

Heisenberg helped to create this orthodox interpretation of the quantum formalism, and accepted it. But he was eventually willing to discuss *also* the problem of "what happens 'really'".[5] According to his later idea, a wave function represents real "objective tendencies" for the occurrence of a "transition from the 'possible' to the 'actual'". He further asserts—and this is crucial—that these transitions, which I call Heisenberg events, occur at the level of the macroscopic measuring device, which is part of the external world;

> it is not connected with the act of registration of the result in the mind of the observer.

Since he also affirms that the wave function of *orthodox* quantum theory *does* change with the registration of the result in the mind of the observer, it apparently follows that there is, in Heisenberg's later view, not only the *subjective wave function* of the orthodox interpretation, but also an *objective wave function*, which exists outside the minds of men. It represents objective tendencies, and "collapses" with the occurrence of a Heisenberg event. Thus there is, in this view, a close parallel between the representation of "our knowledge" provided by the orthodox theory and the form of external reality itself.

6.4 The Mind–Matter Problem

Perhaps the central thesis of James's monumental text, *The Principles of Psychology*,[6] was that each conscious thought is essentially a complex whole: each thought has components, which can be examined by subsequent analysis, but, *as given*, is a unified whole that cannot be reduced to a collection of parts without destroying its essence. On the other hand, matter, according to the science of James's day, was reducible to a collection of simple local parts. Consequently, there was no possibility of finding within matter, as classically conceived, any faithful image of a human thought. James himself drew from this structural mismatch the conclusion that the classical conception of nature was essentially deficient: he apparently anticipated important developments in the natural sciences that would bring our conception of matter into better alignment with the characteristics of thoughts. However, this prescient idea, that classical physics was fundamentally flawed, was not shared by the psychologists of the early twentieth century, who, instead, recoiled from the entire approach based on introspective analysis of thoughts, and the problem of mind and brain, and embraced the opposed ideology of behaviorism.

True to James's expectation there have been fundamental conceptual changes in physics. Heisenberg's picture of nature envisages events that actualize, as units, entire patterns of action in the material world. This physical process permits the emergence, in human brains, of holistic structures that can mirror, simultaneously, both the *structural forms* and *functional effects* of conscious human thoughts.

6.5 Mental Events as Heisenberg Events

According to William James our conscious thoughts have an eventlike quality: they appear as "buds" of reality—either all or nothing at all. This holistic "all or nothing at all" property is precisely the characteristic feature of quantum phenomena that distinguishes them from classical phenomena: either the entire mark appears on the photographic plate, or no mark appears at all; either the entire Geiger counter suddenly discharges, or there is no discharge at all; either the pointer on the measuring device swings suddenly to the right, or it swings suddenly to the left—there is no intermediate possibility. In Heisenberg's picture of nature these discrete events occur in "measuring" devices, which amplify small-scale changes into large-scale signals.

Brains are similar in this respect to measuring devices. At the synaptic events, and also at neuron firings, there are large amplification effects. Moreover, and this is the crucial point, an analysis by John von Neumann[7] shows that the quantum events in the brain need not occur either at the level of the individual synaptic discharge or at the level of the individual neuron-firing: they can occur, instead, at the level of the entire brain, in conjunction with the eventlike occurrence of a conscious thought. Such a quantum event in the brain would actualize, or be the actualization of, an entire complex, quasi-stable, large-scale pattern of neural firings. An actualized pattern of this kind can have the complexity and causal efficacy needed to represent both the structural form and functional effect of a conscious thought.

The basic postulate of this understanding of the mind–brain connection, as I have developed it in reference 8, is that conscious thoughts are represented within the physicist's description of nature by Heisenberg events that actualize entire complex patterns of neurological activity. It is shown how, at least in principle, this postulate can lead to a relatively simple, but scientifically adequate, conceptualization of the mind–matter connection. Some chief features of the theory are that it provides for:

1 An isomorphic connection (i.e., a one-to-one mapping that preserves certain structural relationships) between the structural forms of conscious thoughts, as described by psychologists, and corresponding actualized structural forms in the neurological patterns of brain activity, as suggested by brain scientists.
2 A correspondence at the functional level between the experiential event and the corresponding Heisenberg event in the brain.
3 An explanation of puzzling temporal anomalies observed in mind/ brain research (e.g., in the Libet experiments).
4 An explanation of puzzling temporal anomalies observed in psychophysical experiments. (e.g., in the Kolers–Grünau experiment).
5 A mechanical explanation of the efficacy of conscious thoughts.
6 An explanation of the eventlike and holistic natures of conscious thoughts.

This psychophysical theory, as presently conceived, does not seek to go beyond the Heisenberg picture of nature: it merely extends that idea from external measuring devices to human brains. At this level, the theory is apparently compatible with the Churchland thesis that "mind is brain",[9] or at least an *aspect* of brain. For, according to the theory, every structural and functional aspect of each conscious thought is completely represented within the physicist's representation of the host brain. Even the special "feel" of a conscious thought, the feel that the thought is somehow *bringing itself into being,* is captured by the actualizing quality of Heisenberg's quantum transition. It is hard to see how a theory could do any better job of representing a conscious thought, and the theory that does it is precisely a quantum-theoretical picture of what is happening to the *brain.* Psychodynamics has become an aspect of brain dynamics.

This unification of our understanding of the physical and psychological aspects of nature resolves a long-standing problem. However, it does not address the question of *agency*: it does not say what *causes* the quantum transition to occur? Are these happenings indeed purely random, or are they controlled by some still-hidden level of reality?

6.6 Comparisons with the Ideas of Pauli

Pauli's idea of a regulative principle lying beyond the mind–matter distinction is intertwined with the Jungian concepts of archetype and synchronicity. Synchronicity refers to the occurrence of representations of archetypes *in meaningful coincidences that defy causal explanation.* Pauli apparently believed, perhaps on the basis of his own experience, that the synchronistic aspects of nature identified by Jung were sufficiently striking to place them beyond the bounds of explanation in terms of pure chance. This judgement, if correct, would mean that behind the processes of nature that we already know and understand there lies another, which *acausally* weaves *meaning* into the fabric of nature.

In classical physics the course of events is determined in a local-mechanical way: the change of each local part is fixed by its immediate environment. Thus nature is "myopic": every aspect of the orderly evolution is governed by "perceptions" that have no breadth of vision at all; only infinitesimally nearby things can have any influence. But "meaning" has to do with a grasping of wholes, and this demands expanded vision. Consequently, there is no way for nature's process, as it is understood in classical physics, to incorporate or embody meaning.

In quantum physics the situation seems even worse. The basic process first generates probabilities, or tendencies, by a similar sort of senseless local process. Any residual hope for meaningfulness is then dashed by the entry of pure chance: the actually occurring course of events is fixed by the blind rolls of cosmic dice.

Yet it is an absurdity to believe that the quantum choices can appear simply randomly "out of the blue", on the basis of absolutely nothing at all. *Something* must select which of the possible events actually occur. If this something is a local-mechanical process then we are back again at square one. However, the quantum selection process, if it exists at all, *cannot be local*: it must, at least in certain circumstances, allow happenings in one region to be influenced by human choices made in a distant region at essentially the same time.[10] Moreover, according to the psychophysical theory described above, this process actualizes *intrinsic wholes* that are simultaneously thought and matter. These two features make the underlying quantum process strikingly similar to Jung's synchronistic process: both are *acausal*, i.e., violate the principle of local causation, and both manifest holistic structures in a realm that lies, in the words of Pauli, "beyond the difference between the 'physical' and the 'psychical'".

Logical arguments imply, as just mentioned, that the underlying quantum process must involve instantaneous influences. These would be expected to produce *causal anomalies*, i.e., observed coincidences that cannot be explained in terms of normal ideas of causality. However, the structure of quantum theory guarantees that all traces of these peculiar influences must disappear from the *statistical averages* that occur in the empirical scientific tests: all acausal aspects are completely masked by the effects of chance. This masking depends crucially, however, upon the exact validity of the quantum-statistical rules: if the probabilities specified by quantum theory were to disagree with those defined by nature herself, then the way would be opened for the appearance of causal anomalies.

This interlocking of causality and chance has important consequences. It means that the play of quantum chance acts both to veil the form of fundamental reality and to unveil the form of empirical reality. However, if causal anomalies actually do appear then the veil has apparently been pushed aside: we have been offered a glimpse of the deeper reality.

My interpretation of Pauli's interest in Jungian ideas, in connection with the development of science, is precisely that he saw in the acausal character of the phenomena studied by Jung some evidence of a breakdown in the quantum laws of chance, and, hence, a possible opening to the deep question that quantum theory fails to address—and that Bohr's philosophy encourages us to ignore—namely the question of what decides what actually happens.

What is salient here is that Jung purports to construct, *on the basis of empirical data*, an idea of the form of the synchronistic process. If he has indeed arrived in this way at some knowledge of nature's process itself, not merely an illusion based on improperly evaluated data, and if the quantum and the synchronistic processes are indeed essentially the same process, then an empirical window may have been opened on the process that had been thought by quantum theorists to lie beyond the ken of empirical knowledge.

Of course, the data involved here are not data of the kind that most physicists are comfortable with. Yet there is no reason why the critical scientific tenor of mind cannot be exercised in domains lying far from physics. (Indeed, Pauli must have done this.) Thus we are led to conceive of science as a large array of scientifically conducted endeavors that covers a very wide range of subjects. The problem then arises of choosing the theoretical foundations for each of these separate parts. If this problem is addressed from within the confines of the individual field, then the number of possibilities is great, and ambiguities and divergences of opinion generally emerge.

Traditionally, scientists cope with such conflicts by insisting on the unity of science: they require the theoretical foundations of the various parts to fit into a single unified framework. This effectively adds to each isolated discipline the extra condition that it fit smoothly onto its neighbors.

The ideal of the unity of science might seem so secure and reasonable as to need no mention. However, if "science" is supposed to cover all of the physical, biological, and psychological sciences, then the demand for unity creates a problem. It can be imposed only if there is a single unified framework into which the foundations of each of these diverse components fit. Our historical inability to find such a framework is the cause of the present fragmentation of science. Yet the conception of nature described above, which is based upon my interpretations of the ideas of James, Heisenberg, and Pauli, does appear to accommodate in a coherent and unified way the bodies of knowledge arising from the fields of classical physics, quantum physics and chemistry, the brain sciences, psychophysical experimentation, introspective psychology, and Jungian psychology. It appears, therefore, to be a viable candidate for the framework that is needed if we are to achieve the ideal of the unity of science.

References

1 W. Pauli, quoted by K. V. Laurikainen, *Gesnerus* **41**, 213 (1984).
2 Ref. 1.
3 I. Newton, *Principia Mathematica, General Scholium* (see I. B. Cohen, *Introduction to Newton's* Principia (Harvard University Press, Cambridge, MA, 1978)); letter to Bentley (1691).
4 N. Bohr, *Essays 1958/1962 on Atomic Physics and Human Knowledge* (Wiley, New York, 1963), p. 60.
5 W. Heisenberg, *Physics and Philosophy* (Harper and Row, New York, 1958), chap. III.
6 W. James, *The Principles of Psychology* (Dover, New York, 1950; reprint of 1890 text).

7 J. von Neumann, *Mathematical Foundations of Quantum Theory* (Princeton University Press, Princeton, 1955).

8 H. P. Stapp, A Quantum Theory of the Mind–Brain Interface, Lawrence Berkeley Laboratory Report LBL-28574 Expanded, University of California, Berkeley, 1991, and chap. 5 of the present book.

9 P. S. Churchland, *Neurophilosophy: Toward a Unified Theory of The Mind/Brain* (MIT Press/A Bradford Book, Cambridge, MA, 1986).

10 J. S. Bell, *Physics* **1**, 195 (1964); H. P. Stapp, Noise-Induced Reduction of Wave Packets and Faster-Than-Light Influences, to be published in *Phys. Rev. A*; Significance of an Experiment of the Greenberger–Horne–Zeilinger Kind, submitted to *Phys. Rev. A*; EPR and Bell's Theorem: A Critical Review, *Found. Phys.* **21**, 1 (1991).

7 Choice and Meaning
in the Quantum Universe

7.1 Choice

How does the world come to be just what it is, and not something else? Classical physics offers only a partial answer. It says that the deterministic laws of nature fix everything over all of spacetime in terms of things at a single instant of time. But the remaining question is then: What fixes things at this single instant of time? What determines the initial conditions?

Classical physics provides no answer at all to this question, or only the equivalent answer "God", where God is the name of whatever it is that fixes those things that are not fixed by the laws of nature, as they are currently understood by scientists. I shall call by the name "choice" any fixing of something that is left free by the laws of nature, as they are currently understood.

Classical physics is not the ultimate scientific theory. It fails at the level of atomic phenomena, and has been replaced by quantum theory. However, the quantum laws, unlike the classical laws, are *indeterministic*: they fix not what actually happens, but only the *probabilities* for the various things that might happen. That is, quantum theory, in its orthodox form, provides no answer to the further question: What fixes what actually does happen?

Physicists have proposed four fundamentally different answers to this question. In the first part of this talk I shall describe these four possibilities. However, one thing is immediately clear. If, at the deepest level, the laws of nature are basically indeterministic, like the laws of quantum theory, then, by definition, choices are not confined to the beginning of time: they must occur under more general conditions. In this case a central question in man's search for an understanding of nature, and his place within it, must be this: Under what conditions are choices made, and what role, if any, do human beings play in the generation of choice?

7.2 Bohr's Approach

The first of the four proposals concerning choice is agnostic: it declines to address the issue of where choices occur, on the grounds that this question does not lie within the province of science, or at least within the province of physics. This is the approach of Niels Bohr, whose general orientation is characterized by the following quotations:

> The task of science is both to extend the range of our experience and reduce it to order . . .[1]

> In physics . . . our problem consists in the coordination of our experience of the external world . . .[2]

> In our description of nature the purpose is not to disclose the real essence of phenomena but only to track down as far as possible relations between the multifold aspects of our experience.[3]

As regards the quantum formalism itself Bohr says:

> The sole aim [of the quantum formalism] is the comprehension of observations obtained under experimental conditions described by simple physical concepts.[4]

> Strictly speaking, the mathematical formalism of quantum mechanics and electrodynamics merely offers rules of calculation for the deduction of expectations about observations obtained under well-defined experimental conditions specified by classical physical concepts.[5]

The quantum formalism referred to by Bohr works in the following way:[6] Let A represent a description in terms of classical concepts of the preparation of an atomic system—i.e., a description in terms of the concepts of classical physics of the construction and placement of the preparing devices. Let B represent a description in terms of classical concepts of a possible response of the detection system—i.e., a set of specifications that will allow technically trained observers to determine whether an observed response lies in the specified class B. Then the basic assumption of quantum theory is that, under appropriate conditions, there are mappings

$$A \rightarrow |\psi_A\rangle\langle\psi_A| \equiv \rho_A$$

and

$$B \rightarrow |\psi_B\rangle\langle\psi_B| \equiv \delta_B,$$

from classical descriptions to operators in a Hilbert space, such that the probability that a result meeting specifications B will occur under the conditions A is given by the formula

$$P[A : B] = |\langle \psi_B | \psi_A \rangle|^2$$
$$= \langle \psi_B | \psi_A \rangle \langle \psi_A | \psi_B \rangle$$
$$= \text{tr} \, \rho_A \delta_B.$$

In accordance with Bohr's precepts, this formalism is nothing but a set of rules for computing expectations pertaining to observations obtained under conditions specified in terms of classical concepts.

Bohr claimed that predictions computed essentially in this way provide all of the confirmable predictions about atomic phenomena that are possible in principle, and that quantum theory provides, therefore, a complete description of atomic phenomena: no theory based on some purported "more detailed" description of the atoms can ever, according to Bohr, yield additional confirmable predictions about phenomena of this kind.

7.3 Everett's Approach

Bohr claimed that the description of atomic (and perhaps subatomic) systems in terms of quantum states is complete. Since the physical universe is composed, in some sense, of atomic (and subatomic) particles it seems reasonable to try to represent the entire universe in the same way that one represents a collection of atoms, namely by an operator in a Hilbert space. However, in doing so it is important to recognize that most of the degrees of freedom referred to by such an operator represent properties that are extremely ephemeral: they are properties that are not directly observable by human beings, and are extremely fleeting on the timescale of human experience. The full universe consists therefore of an exceedingly thin veneer of relatively sluggish, directly observable properties resting on a vast ocean of rapidly fluctuating unobservable ones.

If one examines, theoretically, the evolution of the universe under the assumption that nature's process is governed exclusively by a Schrödinger equation, which is the normal quantum law of evolution, then the following picture emerges: owing to the local character of interactions between particles the properties of nature in the thin veneer of local observable properties are continually splitting into a *statistical mixture* of classical worlds of the kind we observe. By a "statistical mixture" I mean a collection of possibilities each having a definite statistical weight, where this statistical weight can be interpreted as the probability that this particular possibility will be the one that is actually realized in nature.

The proposal of Heisenberg and Dirac, which will be described later, asserts that nature singles out and actualizes one observable branch from among the emerging set of possible ones. Everett's counterproposal is that no such choice is ever made, but that rather the character of human consciousness is such that each individual realm of human experience can accomodate only a single one of these branches, even though all the branches exist together in the fullness of nature. Thus in Everett's picture of nature only one choice need ever be made, namely the choice of the initial state of the universe. This initial state could be taken to be some featureless state, on the grounds of a lack of sufficient reason for any specific feature. Then the particularness of the perceived universe observed by any individual person would not be a reflection of any corresponding particularness of the initial state of the universe: it would *not* be, as in classical physics, merely a transformed expression of the particularness present already at earlier times. Rather the observed particularness would be the particularness of one individual branch of the universe. This branch is generated out of a "quantum soup" by the deterministic laws of quantum evolution, with no intervention of choice.

Everett's proposal[7] has, for physicists, the attraction that it makes quantum theory complete in principle. The theory would, if valid, cover, in principle, not only atomic phenomena but also biological and cosmological processes, for example. However, even the proponents of Everett's theory emphasize that the technical details of this interpretation need to be spelled out in more detail. The problem, basically, is the clash between the continuous character of the description of nature provided by the quantum state and the discrete character of human experience. The Everett universe at the observable level probably does not separate into well-defined discrete branches. The various "branches" appear to blend continuously into each other, owing to the basically continuous character of the elementary scattering and decay processes. In the standard applications of the quantum formalism to atomic phenomena a human agent plays a crucial role of setting up specifications for identifying particular classess of physical events. But in Everett's quantum world the human observers and their devices tend to become amorphous distributions of properties. Consequently, no sharp separation of the observable aspects of nature into discrete well-defined branches has yet been demonstrated. This leaves the technical viability of Everett's proposal open to serious doubt.

This problem of the reconciliation of the discreteness of the perceived world with the amorphous character of its purely quantum description is cleanly resolved by the proposal of David Bohm.

7.4 Bohm's Pilot-Wave Proposal

The quantum formalism is fundamentally statistical in character. Hence it is reasonable to postulate the existence in nature of the actual things that the quantum probabilities are probabilities of. These things will then specify what *actually* occurs.

David Bohm has constructed a model of this kind.[8] In his model there is an ordinary classical world of the kind described in classical physics, and, in addition, also a quantum state. This state is supposed to exist as a physically real thing, not merely as an idea in the minds of scientists. It specifies an extra force that acts on each of the particles of the classical world, and causes them to behave in a way compatible with the statistical predictions of quantum theory.

Bohm's model is simple and instructive. It shows that we need not cling to the idea, advanced by the founders of quantum theory, that nature cannot be described in a thoroughly comprehensible way in terms of properties that are always well defined and that evolve in accordance with well-defined deterministic laws.

Bohm's model does violate one of the basic precepts of classical physics: the force on a particle located at a point generally depends strongly upon the precise locations, *at that very instant*, of many other particles in the universe. This instantaneous connection contradicts the idea of classical relativistic physics that no influence can act over a spacelike interval—i.e., faster than light. On the other hand, a now-famous theorem due to John Bell[9] shows that no deterministic theory of this general kind can exclude faster-than-light influences, if it is to reproduce the predictions of quantum theory. Bell's result can be extended also to indeterministic theories.[10] Thus this nonlocal feature ought not be regarded as objectionable, provided all the *observable* properties conform to relativistic principles, as they indeed do in Bohm's relativistic model.

Bohm's model does however retain one feature of classical physics that can be regarded as objectionable, at least aesthetically. This is the need for an arbitrary-looking choice of initial conditions. In particular, some definite initial position for each of the particles in the universe must be chosen. The idea of such an immensely detailed choice suddenly emerging out of nothing at all seems utterly unreasonable. In the alternative proposal of Heisenberg and Dirac, to be described next, the choices are distributed over space and time, and each choice is made within a specific physical context.

7.5 The Heisenberg–Dirac Proposal

The picture of nature most nearly in line with quantum theory as it is used in practice is that of Heisenberg and Dirac. Heisenberg says:

> The observation itself changes the probability function discontinuously; it selects of all possible events the actual one that has taken place . . . the transition from the "possible" to the "actual" takes place during the act of observation. If we want to describe what happens in an atomic event, we have to realize that the word "happens" can apply only to the observation, not to the state of affairs between two observations. It applies to the physical, not the psychic act of observation, and we may say that the transition from the "possible" to the "actual" takes place as soon as the interaction of the object with the measuring device, and thereby with the rest of the world, has come into play, it is not connected with the act of registration of the result in the mind of the observer.[11]

Heisenberg distinguishes what is actually happening in the physical world from representations of the physical situation in the minds of scientists. Strictly speaking, the quantum formalism pertains exclusively to the latter. However, the extreme precision of the predictions of quantum theory justifies our trying to think of nature herself as represented by a quantum state, which, however, must undergo a sudden "quantum jump" in connection with each selection of an actual result from among the ones previously possible.

Dirac espouses a similar idea when he speaks of a "choice" on the part of nature.

The intervention of "choice" in the proposal of Heisenberg/Dirac is completely different from this intervention in the proposals of Everett and of Bohm. In Everett's model there need be no choice at all, except perhaps a choice of a featureless initial condition: all of the particularness that we observe in nature can be supposed to exist in a single branch that is generated in a completely deterministic way by deterministic laws of motion, but then mistakenly perceived to be the whole of nature by virtue of a limitation in the capacity of each individual human consciousness. In Bohm's model, on the other hand, all choice is confined to a single stupendous choice that can be conceived to be made at "the beginning of time", or at some time in the far distant past. In the Heisenberg/Dirac proposal the choice of initial conditions can be, as in the Everett model, the choice of a featureless state. Then, over the course of time, *choices are made* that inject into the universe the particularness that we observe. Each choice in the present era is taken to be a choice from among the observable possible branches that are generated by the deterministic laws of quantum evolution. Under the condition that prior choices have been made, this process can be conceived to generate, at the level of local observable properties, a statistical mixture of reasonably distinct branches, some *one* of which will be selected.

The brain of an alert human observer is similar in an important way to a quantum detecting device: it can amplify small signals to large macroscopic effects. The Heisenberg/Dirac proposal, if taken seriously, must therefore be expected to entail quantum events in the brain that are analogous to the events that are postulated to occur in quantum detecting devices. On the other hand, a quantum event in the brain, if it occurs at the level of the entire brain, or a large part of it, could be incomparably more complex than the actualized state of a simple quantum detection device, simply because of the immensely greater complexity of the brain itself, as contrasted to a quantum measuring device.

Suppose the actualized state of the brain is really *actualized*. What can this mean? One possibility is that some characteristic feature of this state becomes an actual "experience". Such a physical feature, if correctly identified, could become the basis of the correspondence between the physical world described by the physicist and neurophysiologist, and the psychic world described by the psychologist. I shall return to this question after a consideration of the nature of meaning.

7.6 Meaning

The idea of meaning entails a sense of direction: a sense of endurance with refinement; a notion of a process that sustains and refines itself. Thus meaning demands mechanism: it demands a machinery that allows a form to be re-created in refined form. Endurance and reproducibility are essential: the form must endure long enough to activate and guide the machinery that sustains and refines it.

States characterized by local observable properties have the required characteristics of endurance and reproducibility, whereas superpositions of such states do not: the interaction of these latter states with their environments quickly destroys the phase connections that define them, and they are consequently unable to reproduce themselves. Thus local observable properties, or properties similar to them, are the natural, and perhaps exclusive, carriers of meaning within the quantum universe.

From this point of view the quantum universe tends to create meaning: the quantum law of evolution continuously creates a vast ensemble of forms that can act as carriers of meaning; it generates a profusion of forms that have the capacity to sustain and refine themselves.

There are among the full set of quantum states that conceivably could be actualized a plethora of possibilities. Yet if we accept the ideas of Heisenberg and Dirac, or the direct evidence of our senses, the forms that actually are chosen are forms of an exceedingly special kind: they are forms that sustain themselves: the pointer on the measuring instrument swings to the right, and this form endures, not in an absolutely static state, but in a state that sustains an enduring semblance of itself.

This essential characteristic of the quantum event is shared by the only things we really know to be actual: our own experiences: each human experience is a form that actualizes itself as an enduring structure.

In a certain sense this property of the actualized forms is logically required. Consider a thousand dots arranged in a small square. Each of the conceivable possible arrangements constitutes a definite form. However, each of these forms is, at the purely intrinsic level, equivalent to every other one: there is no intrinsic distinction between them. Each one is different, but they are all intrinsically equivalent. To specify some significant difference one must go beyond the immediate intrinsic form itself.

Scientists, in their search for simplicity, endeavor to consider the physical universe as self-contained; as not requiring the intervention of some outside agent. To achieve such an end any distinction made by nature between conceivable possible forms must be based on properties intrinsic to the quantum universe itself. One way to draw such a distinction is to consider each form on the basis of what it *does*, or *produces*, in the quantum universe, rather than on the basis of what it is.

If this strategy is adopted then there is one logical distinction between forms that stands out from all others, in the sense that it does not appeal to any structure that lies outside the form itself. This is the property of a form to sustain itself.

From this point of view the proposal of Heisenberg and Dirac can be characterized in this way: the quantum choices are meaningful choices, where "meaningful" is defined intrinsically, within the quantum system itself, without reference to any external criterion of meaning, in terms of sustainability. Each quantum choice pulls itself out of the quantum soup "by its bootstraps"; it justifies itself by the meaning inherent in the sustainability of the form that is actualized. The "meaning" of this choice is, then, not based upon anything lying outside the chosen form: it resides in the sustainability of that form itself.

This introduction of a notion of intrinsic meaning at the level of the elementary quantum event provides the rudiment of a general quantum conception of meaning based on the intrinsic criterion of sustainability.

Within the quantum formalism each Heisenberg/Dirac quantum choice is a grasping, as a unified whole, of a certain combination of possibilities that hang together as a local enduring form. The actualization of this form utilizes, and restructures, some of the quantum potentialities, and produces an immediate rearrangement of the possibilities available for the next event. The specific form of this rearrangement is fixed by the mathematics of quantum theory.

A principal feature of this rearrangement of possibilities is that a choice made in one region instantly affects the possibilities available in far away regions. If the potentiality for a particle to be detected in one detector is actualized, then the potentiality for this particle to be detected in a far-away region immediately vanishes. Thus the quantum choice is, on the one hand, a local affair, because it actualizes a particular meaningful form in a local region of spacetime. On the other hand, the bookkeeping system is global: an adjustment of possibilities is immediately made over the entire spacetime manifold. Thus the basic process of choice is fundamentally global, but it creates locally defined meaning.

7.7 Ramifications

The foregoing discussion of meaning offers something that science is expected ultimately to provide, and that is desperately needed today, namely the basis of a *Weltanschauung*, or world view, that is fully compatible with the available scientific evidence, and which counters the corrosive mechanical world view that arose from the basically incorrect concepts of classical physics. This quantum conception of nature has emerged directly and naturally out of the idea of the quantum world that generally prevails today in the minds of practicing quantum physicists: it rests on the idea of Heisenberg and Dirac that *under particular kinds of conditions*, nature makes a choice. It is based on an examination of the nature of those conditions. The condition under which nature acts was construed as an expression of a criterion of natural value.

It is possible that this criterion of value in natural process applies only at the level of measuring devices. However, it is at least conceivable that the same criterion applies also on other scales, and could be detected as a biasing of quantum choices in favor of the creation of sustainable forms on all levels. Such a biasing should be detectable under laboratory conditions, and may eventually become necessary to introduce into the domains of biology and cosmology, since the ubiquitous existence of sustainable form on all scales may otherwise be impossible to explain in a natural way.

Another possible ramification pertains to the interface between the brain sciences and psychology. It is evident that mental processes are connected in some way to brain processes. However, the nature of the connection is unclear. Indeed, when viewed from the perspective of classical physics such a connection appears totally incomprehensible. For classical physics is fundamentally reductionistic: each macroscopic system is conceived to be *nothing more* than a simple collection of its microscopic parts, each of which is supposed to react in a completely mechanical way to the instantaneous force that acts upon it. On the other hand, each human experience evidently corresponds to an "entire enduring complex macroscopic form" in a human brain. According to the concepts of classical physics, no such physical form can exist as a *fundamentally* unified entity: no such form can exist *except* as a simple collection of its *fundamentally independent* microscopic parts. The fundamentally unified complex conscious thought has therefore, within the classical conception of nature, *nothing of like kind* to which it can correspond. Moreover, if some nonphysical process of "perceiving certain features of the brain as a complex whole" is added to the classical picture of nature, in order to account for the occurrence and character of human experience, then this process, if it is not to contradict the laws of classical physics, can have no back-reaction or influence upon the course of physical events, which is already completely determined, in terms of the motions of the microscopic realities, by the deterministic laws of motion.

The quantum-mechanical conception of nature is altogether different. In this conception each actual thing is fundamentally the actualization of an entire enduring complex macroscopic form. Those aspects of nature that are described in terms of the simple microscopic parts govern only the *tendencies* for the actualization of such enduring complex forms. The occurrence of such complex forms is therefore neither incidental nor external to the basic dynamical process. On the contrary, the actualization of such forms is the entire object of the dynamics, and it is these forms themselves, not the subordinate microscopic parts, that determine what actually happens.

Within the quantum-mechanical conception of nature human experiences are, as regards their intrinsic structural forms, *similar in kind* to the actualities that evidently play the dominant role in high-level brain dynamics. An analysis[12] of the basic features of high-level brain functioning, and of conscious mental process, reveals that one can in fact postulate an *isomorphism* between the intrinsic structure of conscious mental events and the intrinsic structure of a certain class of brain events, conceived of as quantum events. Conscious mental events thereby become naturally correlated with events in human brains, as they are described in the language of quantum theory. The occurrence or nonoccurrence of such brain events is not, however, predetermined by the known laws of physics: such decisions are matters of choice.

7.8 Summary

If an important task of science is to provide man with the empirical foundation of a philosophically satisfactory comprehension of the universe, and his place within it, then classical physics is profoundly deficient in two important ways. The first concerns choice and meaning. If a "choice" is defined to be a fixing of an aspect of the universe that is not fixed by the known laws of nature, then at the stage of classical physics all choice is confined to "the beginning of time": all choice is compressed into some stupendous initial act, which arises out of nothing at all, or at least out of nothing representable within the physical theory. The universe is consequently rendered "meaningless" from the perspective of man, because each human being is reduced to a mechanical automaton whose every action was preordained prior to his own existence.

The advance to quantum theory appears at first to offer no basis for any significant improvement: choice is now distributed over time, and is confined to particular kinds of physical contexts, but is asserted to be controlled exclusively by "pure chance". Thus we are presented with the two horns of the dilemma, "determinism" or "chance": neither option appears to offer any possibility for a meaningful universe, or a meaningful role for man within it.

Closer study, however, reveals quite the opposite. An examination of *the conditions under which quantum choices are made*, according to the "orthodox" ideas of Heisenberg and Dirac, shows that, even though these choices are not fixed by the quantum laws, nonetheless, each such choice is *intrinsically meaningful*: each quantum choice injects meaning, in the form of enduring structure, into the physical universe.

The second profound deficiency of classical physics is its essentially reductionistic character. According to the concepts of classical physics each thing is essentially nothing more than a sum of simple parts. But this limitation excludes the possibility of the existence, within the physical universe itself, of a faithful representation of a *comprehension* of anything; of a representation within the physical universe of anything that mirrors the essential attribute of a conscious thought, namely its existence as a fundamentally complex whole. The fundamental characteristic of a comprehension, or a thought, is precisely that it is *more* than the sum of its component parts: *it cannot be analyzed into nothing more than the sum of its components without eliminating its very essence*. Thus within the physical universe, as classically conceived, there is no possibility of representing a comprehension of anything: one is forced to look outside the classically conceived physical universe to locate human thoughts. On the other hand, it is the essence of Heisenberg/Dirac quantum events that they choose, and actualize within the physical universe itself, as quantum-mechanically conceived, complex meaningful wholes. Science thus provides man with at least the rudiments of a cohesive view of nature in which his own thoughts and actions are integral parts of a universe that generates meaningful options via the laws of nature, but is not rigidly controlled by these laws.

References

1 N. Bohr, *Atomic Theory and the Description of Nature* (Cambridge University Press, Cambridge, 1934), p. 1.
2 Ref. 1, p. 1.
3 Ref. 1, p. 18.
4 N. Bohr, *Atomic Physics and Human Knowledge* (Wiley, New York, 1958), p. 90.
5 N. Bohr, *Essays 1958/1962 on Atomic Physics and Human Knowledge* (Wiley, New York, 1963), p. 60.
6 H. P. Stapp, The Copenhagen Interpretation, *Am. J. Phys.* **40**, 1098 (1972) and chap. 3 of the present book.
7 H. Everett III, *Rev. Mod. Phys.* **29**, 463 (1957).
8 D. Bohm, *Phys. Rev.* **85**, 166, 180 (1952).
9 J. S. Bell, On the Einstein–Podolsky–Rosen Paradox, *Physics* **1**, 195 (1964).
10 H. P. Stapp, Significance of an Experiment of the Greenberger–Horne–Zeilenger Kind, scheduled for publication in *Phys. Rev. A* (February 1993).
11 W. Heisenberg, *Physics and Philosophy* (Harper and Row, New York, 1958), chap. III.
12 H. P. Stapp, A Quantum Theory of the Mind–Brain Interface, Lawrence Berkeley Laboratory Report LBL-28574 Expanded, University of California, Berkeley, 1991, and chap. 5 of the present book.

8 Future Achievements
to Be Gained through Science

8.1 Introduction

The ideas to be developed here are related to those of two of the other panelists, and it will be useful to describe the connections right at the outset.

The next section is related to the paper of Edward Teller. The topic assigned to Teller was "The Limitations of Physics". However, his conclusion was an *openness* of the possibilities for science, limited only by the imagination of man, not by anything inherent in the nature of science itself. Teller's paper describes many of the anticipated achievements of science, and it has, in good measure, done my job for me. Indeed, the full content of my talk, as it pertains specifically to achievements of science, is an elaboration of one single point made by Teller.

In his section on understanding, Teller says:

> In order to understand atomic structure, we must accept the idea that the future is uncertain. It is uncertain to the extent that the future is actually created in every part of the world by every atom and every living being.
>   This point of view, which is the complete opposite of machinelike determinism, is something that I believe should be realized by everyone.

What does Teller mean when he says "the future is uncertain to the extent that it is actually being created by . . . every living being?" If, as Teller asserts, it is important for everyone to realize that nature is not like a machine, that we are not mere cogs in a giant machine, then it should also be important for everyone to realize what, instead, we actually are: What is the nature of man? What is his place in the universe? What new information has science provided about these basic questions? These are the issues that will be pursued in the section 9.2.

Section 9.3 is related to the contribution of Roger Masters, and it addresses the issue that is the principal focus of the present series of three conferences. This issue is the question of the rational foundation of human values. The conclusion of the first of the three conferences was that the hopes of the enlightenment have not yet been fulfilled. The underlying question being addressed by the present conference is whether there are any permanent limitations in science that must forever prevent science from providing the methodology and foundation for the construction of an adequate system of values for mankind.

Masters addresses the oft-made claim that science can never pass from Facts to Values, from Is to Ought. He argues toward the conclusion that science, if broadly conceived, faces no such limitation. The third section of the present paper addresses this same question, but on the basis of the quantum-ontological foundations described in section 9.2, and it enlarges Masters's conclusion.

8.2 Post-Cartesian Science

The achievements of science are commonly measured in terms of their technological fallout. However, science, at least in the minds of many scientists, is not primarily an adjunct to engineering practice. It is fundamentally a part of man's unending quest for knowledge about the universe and his place within it. This knowledge can, in due course, become vastly more important than the technologies it spawns. For new technologies can only expand our already immense physical capabilities, whereas new knowledge can influence, on a worldwide scale, the thoughts men think, and, specifically, can shape the values and aspirations that determine the entire direction of the human endeavor. In terms of net impact upon human life the most important impending development in science will be, I believe, ideological, not technological. It will be a profound revision of science's conception of man himself: the emergence of a wholly new scientific image of man and his place in the universe.

The contemporary scientific image of man is essentially the image created by classical mechanics. It is erected upon Descartes's idea that nature is divided into two parts, mind and matter. In this classical view the essence of man, namely his consciousness, is torn from his body and forced to reside, impotently, outside the world described by physicists. Philosophy is incapacitated, for it is impossible to erect a coherent philosophy of man and nature upon this incoherent foundation.

For more than two centuries this split image of man had seemed to be mandated by the overwhelming successes of classical mechanics. However, we now know that even the basic precepts of classical mechanics are profoundly incorrect. This failure of classical mechanics at the foundational level removes all justification for retention in philosophy of Descartes's dualistic conception of man. The successor to classical mechanics, quantum mechanics, allows each man's consciousness to be understood as an integral part of the world described in the mathematical language of physics.

Man's image of himself might seem at first to be some airy philosophical abstraction, of minor importance in the shaping of human events, compared to political and economic realities. However, personal self-image is a primary driving force in human affairs. Now, as throughout history, it is men's image of themselves as agents of emerging spiritual or secular orders that generates the passionate commitments that power the major currents of history.

The first clear sign within science of a turning away from Descartes's dualistic conception of nature appears in the orthodox quantum philosophy of Niels Bohr. According to that philosophy the basic realities in the scientific description of nature are the experiences of human observers. The mathematical structures of physical theory, which in classical mechanics were imagined to represent, in an accurate way, an external world of matter, become recognized as mere tools for the description and prediction of human experiences.[1] Thoughts, which stood impotently outside the real physical world of classical mechanics, became the only accepted realities in Bohr's quantum-mechanical description of nature.

Bohr's quantum philosophy does not pretend to be ontologically coherent: it is concerned with what we can know, not with what "really" exists. On the other hand, Werner Heisenberg, in an effort to provide an understanding of what is actually happening in nature, discussed a model of objective reality itself.[2] Heisenberg's general conception of reality is probably the idea of nature most widely accepted today by quantum physicists.

Heisenberg's idea of reality, like that of Descartes, is based on a separation of nature into two parts. However, these two parts, unlike the two parts of Descartes's universe, are logically inseparable. They correspond to the wavelike and the particlelike aspects of nature respectively — to the two parts of the wave–particle duality.

The wavelike aspect of nature is represented by Heisenberg's "state of the universe": every wavelike feature of nature is embedded in the Heisenberg "state". This state, in conjunction with an associated structure of "operators", gives information pertaining to points located everywhere in space. Moreover, it gives information pertaining to every instant of time, from the infinite past to the infinite future.

Heisenberg's state of the universe, in conjunction with the associated structure of operators, is a quantum analog of the classical description of "matter". It encompasses equations of motion analogous to Newton's equations motion for matter, and represents a vast system of spacetime relationships. However, it does not describe even approximately the nature that appears in human perceptions. Instead, this state has a separate "branch" corresponding to each of the alternative possibilities that we *might* come to observe. To each of these alternative observable possibilities, the state assigns a statistical weight. This statistical weight represents, intuitively, in Heisenberg's words, the "objective tendency" for that possibility to appear.

For example, if we set up a Geiger counter to monitor the radioactivity of a certain sample of radium, then the Heisenberg state of the universe would contain branches corresponding to each of the possible instants of time at which that Geiger counter might fire. Furthermore, it would give, for any specified time interval, the probability that the counter would fire during that interval. However, *every* possible time of firing would be represented in the state; no single time is picked out as the time at which the counter actually fires.

To complete his model of nature Heisenberg adds, therefore, a second part: a sequence of "actual events". Each actual event represents the selection and actualization of a single large-scale (i.e., observable) pattern of activity in some large system, such as a Geiger counter.

The two parts of the Heisenberg model are tightly interwoven: each Heisenberg *event* is completely described as a change of the Heisenberg *state* from its form prior to this event to its form subsequent to the event. Conversely each Heisenberg *state* is nothing but a representation of the tendencies associated with the various alternative possibilities for the next Heisenberg *event*. Thus each part is nonsense without the other: a change of the state can make no sense without states; and tendencies for events has no meaning without events. Moreover, each part of Heisenberg's ontology is securely rooted in physical phenomena: the "states" account for the wavelike aspects of nature, and the "events" account for the particlelike aspects. The two parts of Heisenberg's ontology are therefore, in contrast to the two parts of Descartes's ontology, both logically inseparable and securely tied to mathematically representable features of observed physical phenomena.

Heisenberg's separation of reality into "events" and "states" seems at first completely different from Descartes's separation of reality into "mind" and "matter". A Heisenberg event such as the firing of a Geiger counter is supposed to occur whether or not anybody is looking: consciousness is not involved in the firing of an unobserved Geiger counter. Moreover, a Heisenberg "state" describes only tendencies for events, rather than any independent and persisting "matter", in the sense of classical mechanics. "Events" are needed to complete the picture. Thus Heisenberg's ontology, taken as a whole, seems to replace the "matter" part of Descartes's duality: the "mind" part seems to be still left out.

To see the relevance of Heisenberg's ontology to consciousness, one must apply his ontology to brains. A key point is that each Heisenberg event represents, as already emphasized, the selection and actualization of a *large-scale* pattern of activity in a *large* physical system. For a brain event this large physical system *can perfectly well be the whole brain, or a large part of it.*

Exploitation of this possibility leads to a natural understanding of the connection of consciousness to brain processes. It is assumed that a certain class of Heisenberg events in the brain are the selection and actualization of large-scale patterns of neural excitations that are "facilitated" for later re-excitation. A detailed study based on a substantial amount of scientific evidence from brain sciences, psychology, and quantum physics, indicates that the psychological structure of each human conscious event can be adequately represented by certain specified features in the structure of the large-scale neural patterns of activity selected and actualized in this way. The intrinsic structure of the psychological event, as an element in a psychological realm, becomes identical to an actualized structure inhering in the neural activity, considered as an element in a realm of physiological structures.[3]

The characteristic *ontological* quality of the Heisenberg actual event is its "actualness"; its property of being a "coming into beingness". This property is also the ontological quality of a conscious event. Thus the conscious event and the (specified features of the) brain event are *structurally and ontologically* indistinguishable: these two corresponding events are, within the mathematical theory itself, the same thing. Consciousness is not, therefore, in this quantum-mechanical description of nature, something that hovers outside of space and matter, observing the mathematically described world but not influencing it. Rather, it is representable as an integral natural part of the basic dynamical process that gives form to the universe, and its structure is completely represented within the physicist's mathematical description of nature.

Let me elaborate upon this essential point. In Heisenberg's ontology the basic process in nature is a sequence of actual events. Each of these events selects and actualizes some large-scale pattern of activity in some large physical system. This actualization is represented by a "quantum jump" in the Heisenberg state of the universe: the old state jumps to a new state. This new state specifies the tendencies for the next actual event, and so on. It is this sequence of actual events that creates the evolving form of the universe. The mental life of each human being is representable as a sub-sequence of the full sequence of Heisenberg events.

Each human brain, like every other physical system, consists not of substantive "matter", but rather of "potentia", or "objective tendencies", for the subsequent actual event. Each human conscious event, like every other Heisenberg event, selects and actualizes a large-scale pattern of activity in some large physical system, this system being, in the case of a human conscious event, a human brain. This brain event actualizes a single coherent pattern of neural activity from a collection of patterns that were possible prior to that event.

This dynamical process makes the Heisenberg universe essentially different from the reductionistic/deterministic universe of classical mechanics. Now the basic process of nature is a sequence of acts each of which selects and binds together diverse strands of potentiality into a single actualized spatially extended whole.

Each Heisenberg actual event is, in the sense just described, localized in a particular physical system. However, it is fundamentally global in character: it is a change in the Heisenberg state of the whole universe, and as such it immediately institutes changes in tendencies everywhere in the universe. This nonlocal feature had made this conception of nature seem improbable, prior to a celebrated theorem due to John Bell, which showed, essentially that this sort of instantaneous action is unavoidable (unless the macroscopic world is wildly different from what it appears to be).

The place of human consciousness in this quantum universe is entirely different from the place of human consciousness in the classical universe. No longer is man an isolated and impotent cog in a mindless machine. Rather he is, through his consciousness, an integral part of the global, mindful, integrating process that gives form to the universe.

8.3 Science and Values

It is often maintained that science stands mute on the question of values: that science can help us to achieve what we value once those values are fixed, but can play no role in determining values. That claim is certainly incorrect: what we value depends upon what we believe, and what we believe is increasingly determined by science. Indeed, the contemporary disarray in the field of values was largely caused by science, for science destroyed the credibility of the myths upon which prior value systems were based, but offerred no adequate replacement.

The purported logical "gulf" between scientific facts and human values would constitute, if true, a permanent limitation in science. This gulf was, accordingly, the principal focus of the preceding talk by Roger Masters.

Although the conclusion reached by Masters is in some ways similar to my own, he sees the critical change as a partial return from modern science to ancient science, whereas I see it as a direct advance from classical science to quantum science.

In discussing this issue, terminology turns out to be important. Masters speaks of ancient science, modern science, and postmodern science, whereas I speak of ancient science, classical science, and quantum science, or, equivalently, of pre-Cartesian science, Cartesian science, and post-Cartesian science. My three stages are basically different from his. His "modern science" is defined essentially by the ideas of Francis Bacon, John Locke, and David Hume, whereas my "classical science" is defined essentially by the ideas of Descartes, Galileo, Newton, Maxwell, and Einstein. My quantum, or post-Cartesian, science is defined essentially by the ideas of Heisenberg; it is not similar to the nihilism associated with postmodernism. The ancient science of Masters is built around the ideas of Socrates, Plato, and Aristotle, whereas I would perhaps focus on Pythagoras and Euclid, simply to emphasize the essential role of mathematics in my conception of science.

To appreciate the connections between Masters's words and mine, one must recognize that his "modern science" is quite different from my "classical science". The "conquest of nature" is an essential part of "modern science", whereas the aim of "classical science" is to know nature not to conquer it: engineering is regarded as a separate endeavor.

The "gulf" between fact and value is regarded as an assumption of modern science, in the view of Masters. In classical mechanics this "gulf" is neither an explicit assumption, nor a defining characteristic. It is simply a consequence of the fact that classical mechanics does not deal with the whole of man as a coherent element in the scientific description of nature. Man's consciousness is explicitly omitted from the classical description. So there is no way for classical science to come to grips with the essentially human questions of values. On the other hand, post-Cartesian quantum science brings consciousness directly into the scientific description of nature. Since quantum mechanics is able to describe the whole of man as a coherent aspect of the process of nature, the theory has, at least in principle, some possibility of saying something about human values.

Masters presents the argument for the "gulf" between facts and values by quoting a passage from Arnold Brecht:

> Deductive analytic logic . . . can add nothing to the meaning of propositions; it can merely make explicit what is implied in that meaning. Inferences of what "ought" to be, therefore, can never be derived deductively (analytically) from premises whose meaning is limited to what "is"; they can be correctly made only from statements that have an Ought-meaning, at least in major

premise . . . In logic there is, as some have expressed it, an "unbridgeable gulf" between Is and Ought.[4]

Masters follows this quotation with the words:

In human affairs, the questions of ends or goals need be viewed as moral, religious, or personal values that are distinct from knowledge of fact: science can be understood as establishing "scientific value relativism", a doctrine that can explore the implications of value choices but never provide a rational or scientific ground for those choices themselves.

These arguments buttress the idea that science can provide no logical foundation for values. However, Masters then quotes Hume:

Thus the course of the argument leads us to conclude, that since vice and virtue are not discoverable merely by reason, or by comparison of ideas, it must be by means of some impression, or sentiment they occasion, that we are able to mark the difference betwixt them . . . Morality, therefore, is more properly felt than judged of; though this feeling is so soft and gentle that we are apt to confound it with an idea, according to our common custom of taking all things for the same which have any near resemblance to each other.[5]

Masters follows this with his summary:

Hume derives what we ought to do (values, the virtues) from observable feelings that can be studied by modern science (facts).[6]

It is, of course, true, and essential, that facts elicit feelings, and feelings have motivational impacts; they cause us to act in certain ways: the cold fact that my life is in danger elicits feelings that motivate me to act to save it, because I value my life. However, the basic problem is not what one does in fact value, which is a factual question that can indeed be dealt with by science. It is rather what one "ought" to value. But what is the rational or scientific basis of ought-statements?

The answer is the following rational imperative:

Rationally, I ought to act in accord with my rationally constructed self-image.

Thus the relevance of scientific knowledge to "ought" is via rationally constructed self-image.

The basis of the imperative asserted above is the basic law of mind: mind abhors confusion; it seeks coherence. The value of rational thought is that it banishes inconsistency: it cannot lead to conflicting conclusions, and hence to confusion. This character of mind, that it abhors confusion, may have a biological foundation or function: we think in order to formulate coherent plans of action, and contradiction is inimical to a coherent plan of action. To maintain the freedom from confusion that is the whole purpose of rational thought one must act in accord with a rationally constructed self-image: to act contrary to a rationally constructed self-image is to create the very confusion that mind abhors.

The basic issue in the realm of values is therefore self-image: What does the person believe himself to be? If science and rational thought convince him that he truly is what the quantum-mechanical image of him described above claims him to be, then he must, if he acts rationally, act in accord with that image.

But what sort of behavior accords with this quantum-mechanical image of man?

The behavior consistent with an isolated cog in a mindless machine would to be to act in accordance with the belief that everything is fated, anyway, and that neither "I", nor anyone else, can either do anything about, or be responsible for, anything that happens in the world, even personal voluntary acts. This sort of "rational" view is not uncommon today. On the other hand, the person who recognizes himself to be an integral component of a universal process that selectively weaves waiting potentialities into dynamic new forms that create potentialities for still newer integrations should be inspired to engage actively and energetically in the common endeavor to enhance the creative potentialities of all of us.

References

1 H. P. Stapp, The Copenhagen Interpretation, *Am. J. Phys.* **40**, 1098 (1972) and chap. 3 of the present book.
2 W. Heisenberg, *Physics and Philosophy* (Harper and Row, New York, 1958), chap. III.
3 H. P. Stapp, A Quantum Theory of the Mind–Brain Interface, Lawrence Berkeley Laboratory Report LBL-28574 Expanded, University of California, Berkeley, 1991, and chap. 5 of the present book.
4 A. Brecht, *Political Theory* (Princeton University Press, 1968), pp. 126–27.
5 D. Hume, *Treatise on Human Nature* (Doubleday, Anchor, Garden City, 1961), III, i, 2, p. 424.
6 R. D. Masters, The Assumptions and Techniques of Science: Knowledge, Values, and the Return to Naturalism (contribution to the panel discussion "The Permanent Limitations of Science", the Claremont Institute, Claremont, California, 14–16 February 1991).

9 A Quantum Conception of Man

9.1 Introduction

Science has enlarged tremendously the potential of human life. By augmenting our powers it has lightened the weight of tedious burdens, and opened the way to a full flowering of man's creative capacities. Yet, ironically, it is the shallowness of a conception of man put forth in the name of science that is the cause today of the growing economic, ecological, and moral problems that block that full flowering.

How could a shallow conception of ourselves, a mere idea, be the cause of such deep troubles? The answer is this: Our beliefs about ourselves in relation to the world around us are the roots of our values, and our values determine not only our immediate actions, but also, over the course of time, the form of our society. Our beliefs are increasingly determined by science. Hence it is at least conceivable that what science has been telling us for three hundred years about man and his place in nature could be playing by now an important role in our lives. Let us look at what actually happened.

The seventeenth century was time of momentous change in men's ideas about the world. During that period thinkers like Galileo, Descartes, and Newton transformed the world, as seen by educated men, from a place where spirits and magic could flourish, to a world of machines: the entire universe came to be viewed as a giant machine, running on automatic, with each of us a tiny cog within it. The symbols of the age that followed were the factory, the steam engine, the railroad, and the automobile. Later on, during our own century, this mechanical age would become transformed in turn by thinkers such as Heisenberg, Schrödinger, and Bohr into the quantum age, whose symbols would be not roaring factories but giant transistorized computers, silently bonding all parts of the planet, with men becoming not so much bodily cogs in a giant machine as mental hubs in a burgeoning network of ideas.

The seventeenth-century transition from the medieval to the mechanical age was triggered by a seemingly miniscule change in a single idea: the orbits of the planets were found to be neither circles, nor circles moving on circles, but ellipses. This apparently trivial and recondite detail, discovered by the scientist Johannes Kepler, through laborious analysis of a mass of astronomical data, was the foundation upon which Isaac Newton built modern science, and simultaneously discredited both centuries of philosophical dogmas and the methods of thinking that produced them. Painstaking observation of nature, and analysis of the empirical findings, came to be seen as a truer source of knowledge than pure philosophical reflection. That kind of reflection had led to the notion that, because circles are perfect figures, and everything in the heavens must be perfect, all planets must move on circles, or at least on circles compounded. But Newton's laws decreed that the orbits of planets were ellipses, not epicycles, and the entire empire of medieval thought began to crumble. In its place rose another, based on Newton's idea of the world as machine. Later on, when this mechanical idea gave way in turn to the quantum one, it was again a mass of esoteric data, analyzed to reveal a totally unexpected structure in nature, that combined to overthrow a conception of the world that had become by then an integral part of the fabric of human life.

The focus of our interest here is on the relationship between the mental and material parts of nature. Human beings have an intuitive feeling that their bodies are moved by their thoughts. Thus it is natural for them to imagine that thoughts of some similar kind inhabit heavenly bodies, rivers and streams, and myriads of other moving things. However, the key step in the development of modern science was precisely to banish all thoughtlike things from the physical universe, or at least to limit severely their domain of influence. In particular, Descartes, in the seventeenth century, divided all nature into two parts, a realm of thoughts and a realm of material things, and proposed that the motions of material things were completely unaffected by thoughts throughout most of the universe. The only excepted regions, where thoughts were allowed to affect matter, were small parts of human brains called pineal glands: without this exception there would be no way for human thoughts to influence human bodies. But outside these glands the motions of all material things were supposed to be governed by mathematical laws.

Carrying forward the idea of Descartes, Isaac Newton devised a set of mathematical laws that appeared to describe correctly the motions of both the heavenly bodies and everything on earth. These laws referred only to material things, never to thoughts, and they were complete in the sense that, once the motions of the material parts of the universe during primordial times were fixed, these laws determined exactly the motions of atoms, and all other material things, for the rest of eternity. Although Newton's laws were expressed as rules governing the motions of atoms and other tiny bits of matter, these laws were tested only for large objects, such as planets, cannon balls, and billiard balls, never for atoms themselves.

According to Descartes's original proposal the purely mechanical laws of motion must fail to hold within our pineal glands, in order for our thoughts to be able affect our bodily actions. However, orthodox scientists of the eighteenth and nineteenth centuries, tolerating no exceptions to the laws of physics, held that each atom in a human body, or in any other place, must follow the path fixed by the laws of physics. This rigid enforcement of the physical laws entailed, of course, that men's thoughts could have no effects upon their actions: that each human body, being composed of preprogrammed atoms, is an automaton whose every action was predetermined, long before he was born, by purely mechanical considerations, with no reference at all to thoughts or ideas.

This conclusion, that human beings are preprogrammed automata, may sound absurd. It contradicts our deepest intuition about ourselves, namely that we are free agents. However, science, by pointing to other situations where intuition is faulty, or dead wrong, was able to maintain, on the basis of its demonstrated practical success and logical consistency, that its view of man was in fact the correct one, and that our feeling of freedom is a complete illusion.

This picture of man led, during the eighteenth and nineteenth centuries, to an associated moral system. It was based on the principle that each of us, being nothing but a mechanical device, automatically pursues his calculated self-interests, as measured by a certain bodily physical property, which is experienced in the realm of thought as pleasure. This principle, which was in line with the commercial temper of the times, was fundamentally hedonistic, though, from the scientific viewpoint, realistic. However, philosophers were able to elevate it to a more socially satisfactory idea by arguing that the "enlightened" rational man must act to advance his own "enlightened" self-interest: he must act to advance the general welfare in order to advance, in the end, his own welfare. Yet there remained in the end only one basic human value: no noble, heroic, or altruistic aim could have any value in itself; its value must be rooted in the common currency of personal pleasure. This kind of morality may seem to be immoral but it appears to be the rational outcome of accepting completely the mechanical or materialistic view of man.

This view of man and morals did not go unchallenged. Earlier traditions lost only slowly their grip on the minds of men, and romantic and idealistic philosophies rose to challenge the bondage of the human spirit decreed by science. From the ensuing welter of conflicting claims, each eloquently defended, followed a moral relativism, where every moral viewpoint was seen as based on arbitrary assumptions. This pernicious outcome was a direct consequence of the schism between the mental and material aspects of nature introduced by science. That cleavage, by precluding any fully coherent conception of man in nature, made every possible view incomplete in some respect, and hence vulnerable. In the resulting moral vacuum the lure of material benefits and the increasing authority of science combined to insinuate the materialistic viewpoint ever more strongly into men's thoughts.

This science-based creed contains, however, the seeds of its own destruction. For behind a facade of social concern it preaches material self-aggrandizement. We are now in the thralls of the logical denouement of that preaching. With the accelerating disintegration of the established cultural traditions, brought on by increased fluxes of peoples and ideas, the demand for satisfaction of inflated material desires has spiraled out of control. This has led to a plundering of future generations, both economically and ecologically. We are now beginning to feel the yoke laid upon us by our predecessors, yet are shifting still heavier burdens onto our own successors. This materialist binge cannot be sustained. Yet the doctrine of enlightened self-interest has no rational way to cope with the problem, as long as each human "self" continues to be perceived as a mere bundle of flesh and bones. For if we accept a strictly materialistic way of thinking, then our own pleasure can be enhanced by ignoring calamities that we ourselves will never face.

Men are not base creatures: all history shows them to be capable of elevated deeds. But elevated deeds and aspirations spring from elevated ideas, and today all ideas, if they are long to survive, must stand up to withering scrutiny. They must in the end be rationally coherent, and consistent with the empirical evidence gathered by science. The mechanical ideas of seventeenth-century science provided no rational or intellectual foundation for any elevated conception of man. Yet the ideas of twentieth-century science do. Quantum theory leads naturally to a rationally coherent conception of the whole of man in nature. It is profoundly different from the sundered mechanical picture offered by classical physics. Like any really new idea this quantum conception of man has many roots. It involves deep questions: What is consciousness? What is choice? What is chance? What can science tell us about the role of these things in nature? How does science itself allow us to transcend Newton's legacy? It is to these questions that we now turn.

9.2 Science, Tradition, and Values

This is the third UNESCO Forum for Science and Culture. Our focus throughout the series has been on the interplay of science, tradition, and values in mankind's search for a sustainable future. At the first forum, held in Venice in 1986, the specter of nuclear annihilation loomed as the principal perceived threat to human survival. By the time of the second forum, in Vancouver in 1989, it was the impending disruption of global ecological balances that seemed most critical. Today, in 1992, the nuclear threat may have receded. But the ecological crisis seems to be worsening, and we are faced with problems of socioeconomic collapse: in the former Soviet Union and eastern Europe one of the world's two premier socioeconomic systems has already collapsed, and in the West and the Third World pressures of ethnic rivalries and economic malaise are tending to make many formerly prosperous and stable countries increasingly ungovernable.

Science has been perceived as the major cause of these problems. It gave man the capacity to ignite a nuclear holocaust, to disrupt the ecosystem on a global scale, and to effect swift, massive and untested social and economic changes. At a deeper level of causation, science has revised man's basic idea of himself in relation to nature. In traditional cultures nature was perceived as a mysterious provider, to be revered and deified. But Francis Bacon, herald of science, proclaimed a new gospel for the age of science: man, abetted by science, was to achieve the *conquest of nature*.

At an even deeper level of causation the Cartesian separation between the minds of men and the rest of nature, which was the key to the seventeenth-century scientific revolution, eroded the foundations of moral thought, and left man adrift with no rationally coherent image of himself within nature. He proclaimed himself to be, on the one hand, ruler of nature, yet was, on the other hand, according to the very scientific theories that were to give him dominion, a mere mechanical cog in a giant mindless machine. He was stripped of responsibility for his acts, since each human action was preordained prior to the birth of species, and was reduced to an isolated automaton struggling for survival in a meaningless universe.

In the face of these science-induced difficulties one must ask: Who needs science? What we obviously need is strong remedial action—a curtailing of science-inflated population growth, consumption, waste, and poverty.

But how can the required global actions be brought about? Dire warnings have minimal effects on populations inured to media hype. An immediate disaster at one's doorstep might suffice, but by then full global recovery may be out of reach.

To change human actions globally one must change human beliefs globally. Global beliefs, to the extent they they exist at all, are the beliefs generated by science. However, some of the most important science-generated beliefs that now pervade the world are beliefs that arose from science during the seventeenth, eighteenth and nineteenth centuries, and are now outdated. Twentieth-century science has wrought immense changes in precisely those beliefs that have in large measure created our present problems.

9.3 Science and a New Vision of Nature

Twentieth-century science yields a conception of nature that is profoundly different from the picture provided by the seventeenth century science of Newton, Galileo, and Descartes. Three changes are particularly important.

The first great twentieth-century change is the dethronement of determinism. Determinism is the idea that each stage of the coming into being of the physical universe is completely controlled by what has already come into being. A failure of determinism means that what is happening, or coming into being, at certain stages of the evolutionary process is not completely fixed by what has come before. Those aspects of the evolutionary process that are *not* completely fixed by prior developments can be called "choices" or "decisions". They are in some sense "free", because they are not completely fixed by what has come before.

The second great twentieth-century change is in science's idea of the nature of "matter", or of the "material universe", which I take to be that part of nature that is completely controlled by mathematical laws analogous to the laws of classical physics. The material universe can no longer be conceived to consist simply of tiny objects similar to small billiard balls, or even things essentially like the electric and magnetic fields of classical physics. Opinions of physicists differ on how best to understand what lies behind the phenomena described so accurately by quantum theory. But the idea most widely accepted by quantum physicists is, I believe, the one of Heisenberg. According to this idea the "material universe" consists of none of the things of classical physics. It consists rather of "objective tendencies", or "potentialities". These tendencies are tendencies for the occurrence of "quantum events". It is these quantum events that are considered to be the *actual things* in nature, even though the potentialities are also real in some sense. Each actual event creates a new global pattern of potentialities. Thus the basic process of nature is no longer conceived to be simply a uniform mathematically determined gradual evolution. Rather it consists of an alternating sequence of two very different kinds of processes. The first phase is a mathematically controlled evolution of the potentialities for the next quantum event. This first phase is deterministic, and the laws that control it are closely analogous to the laws of classical physics. The next phase is a quantum event. This event is not, in general, strictly controlled by any known physical law, although collections of events exhibit *statistical* regularities. Thus each individual quantum event creates a new world of potentialities, which then evolves in accordance with certain deterministic mathematical laws. These potentialities define the "tendencies" for the next event, and so on. Each quantum event, because it is not fixed by anything in the physicist's description of prior nature represents a "choice". The critical fact is that each such choice can actualize a *macroscopic* integrated pattern of activity in the newly created material universe of potentialities.

The third great twentieth-century change in science is the recognition of a profound wholeness in nature, of a fundamental inseparability and entanglement of those aspects of nature that have formerly been conceived to be separate. The apparent separateness of ordinary physical objects turns out, in this view of nature, to be a statistical effect that emerges from the multiple actions of many quantum events. It is only at the level of the *individual events* that the underlying wholeness reveals itself.

9.4 Science and a New Vision of Man

The most important consequence of this altered vision of nature is the place it provides for human minds. Consciousness is no longer forced to be an impotent spectator to a mechanically determined flow of physical events. Conscious events can be naturally identified with certain special kinds of quantum events, namely quantum events that create *large-scale integrated patterns of neuronal activity in human brains*. These events represent "choices" that are not strictly controlled by any known physical laws. Each such event in the brain influences the course of subsequent events in the brain, body, and environment through the mechanical propagation of the potentialities created by that event.

This revised idea of man in relation to nature has profound moral implications. In the first place, it shows that the pernicious mechanical idea of man and nature that arose from seventeenth-century science was dependent upon assumptions that no longer rule science.

Contemporary science certainly allows human consciousness to exercise effective top-down control over human brain processes. Hence the idea that man is not responsible for his acts has no longer any basis in science. Moreover, the separateness of man within nature that had formerly seemed to be entailed by science is now reversed. The image of man described above places human consciousness in the inner workings of a nonlocal global process that links the whole universe together in a manner totally foreign to both classical physics and the observations of everyday life. If the world indeed operates in the way suggested by Heisenberg's ontology then we are all integrally connected into some not-yet-fully-understood global process that is actively creating the form of the universe.

The strongest motives of men arise from their perception of themselves in relation to the creative power of the universe. The religious wars of past and recent history give ample evidence that men will gladly sacrifice every material thing, and even their lives, in the name of their convictions on these issues. Thus the quantum-mechanical conception of man described above, infused into the global consciousness, has the capacity to strongly affect men's actions on a global scale.

Science recognizes no authority whose *ex cathedra* pronouncements can be claimed to express a divine will. Nevertheless, this new conception of the universe emphasizes an intricate and profound global wholeness and it gives man's consciousness a creative, dynamical, and integrating role in the intrinsically global process that forms the world around us. This conception of man's place in nature represents a tremendous shift from the idea of man as either conqueror of a mindless nature, or as a helpless piece of protoplasm struggling for survival in a meaningless universe. Just this conceptual shift alone, moving the minds of billions of people empowered by the physical capacities supplied by science, would be a force of tremendous magnitude. Implicit in this conceptual shift in man's perception of his relationship to the rest of nature is the foundation of a new ethics, one that would conceive the "self" of self-interest very broadly, in a way that would include in appropriate measure all life on our planet.

9.5 Discussion

Varela: How does your picture account for the many levels of structure in brain processing that lie between the quantum events at the atomic level and consciousness?

Stapp: In the first place the quantum events are not at the atomic level. According to Heisenberg's idea, the quantum events, that is the actual events, occur only when the interaction between the quantum system and the measuring device, "and hence the rest of the world", comes into play. The actual events that I am talking about occur at a *macroscopic* level: the whole Geiger counter "fires", or the whole pointer on the measuring device is actualized as swinging to the left, rather than to the right. The quantum events select from among the alternative possible *cohesive macroscopic patterns of activity*. As for the many levels of processing in the brain, these are considered to be mechanical *brain* processes: they are consequences of the quantum-mechanical laws of motion, which determine the evolution of the "propensities" for the various alternative possible quantum events. In most other theories of the mind–brain connection there is no basis for a fundamental ontological difference between brain processes that are consciously experienced and those that are not. This is because their basic ontological structure is monistic, rather than dualistic, as it is in quantum theory. Quantum theory thus allows for a *fundamental* physical difference between brain processes that are experienced and those that are not.

Varela: What empirical evidence is there that quantum theory is important in brain processes that are directly connected to consciousness?

Stapp: Chemical processes are essential to brain operation, and hence a quantum description is mandated. In fact, quantum mechanics is essential to any understanding of the properties of materials, be they inorganic, organic, or biological. Classical ideas do not suffice to explain properties of materials, and properties of various materials play an essential role in the functioning of the brain.

Varela: The microscopic atomic properties lead to macroscopic properties, such as electric pulses along neurons, that can be described classically. What empirical evidence is there that a classical description is inadequate for describing those brain processes that are directly connected to conscious process?

Stapp: The processes that can be described classically can also be described quantum mechanically, and the latter description is *fundamentally* better because it fits onto the lower-level chemical processes in a rationally coherent way. Thus one *can* use a quantum description, and at least in principle, *should* use a quantum description, because it is universal, or at least *can be* universal: classical physics is known to be inadequate in some respects: it is known to be nonuniversal.

The quantum description is not only required to explain the underlying atomic and chemical processes, it is fundamentally richer also in the treatment of macroscopic properties, as the theory of consciousness described here shows.

As Quine has emphasized, theories are underdetermined by data. In order to have any hope of achieving a reasonably unique understanding of nature we must insist upon the unity of science, and strive for a coherent understanding that covers the entire range of scientific knowledge. It is only if science can give us a *unified* comprehension of nature that we can turn to it with any confidence for an understanding of our place in nature.

McLaren: You say that a quantum jump selects one of the alternative possibilities, and that this selection is not under the strict control of any known law of nature. And certain of these jumps control the course of brain activity. My question is this: Are not these jumps arbitrary, and if so are we not back in a random universe?

Stapp: These jumps are not strictly controlled by any *known* law of nature. And *contemporary* quantum theory treats these events as random variables, in the sense that only their statistical weights are specified by the theory: the specific actual choice of whether this event or that event occurs is not fixed by contemporary theory.

The fact that contemporary physical theory says nothing more than this does not mean that science will always be so reticent. Many physicists of today claim to believe that it is perfectly possible, and also satisfactory, for there to be choices that simply come out of nowhere at all. I believe such a possibility to be acceptable as an expression of our present state of scientific knowledge, but that science should not rest complacently in that state: it should strive to do better. And in this striving all branches of scientific knowledge ought to be brought into play. There is currently in science a movement toward fragmentation, reflecting the departmentalization of our universities, whereby each discipline within science asserts its autonomy: its right to stand alone as an independent field of study. I believe this movement to be retrograde: that science can succeed in creating a unique plausible picture of all of nature, including ourselves, only by accepting the scientifically established results from all the fields and insisting on a rationally coherent theoretical understanding of all scientifically acquired knowledge. In this broader context the claim that the choice comes out of nowhere at all should be regarded as an admission of contemporary ignorance, not as a satisfactory final word.

Contemporary science certainly *allows* the choices to be other than "purely random". Indeed, in a model of the quantum world devised by David Bohm these choices are deterministically controlled. The basic question, however, is whether there is a rationally coherent possibility that is both compatible with all scientifically acquired data, yet intermediate to these two alternative possibilities of "pure chance" and "pure determinism".

The philosopher A. N. Whitehead speaks of such an intermediate possibility, which is closer to the intuitive idea that our choices are, in some sense, *self-determining*: namely that they are conditioned by what has come before, yet are not strictly determined by the past, but are nonetheless not without sufficient reason. I think such a possibility is open, but to give this logical possibility a nonspeculative foundation will require enlarging the boundaries of scientific knowledge.

10 Quantum Theory
and the Place of Mind in Nature

Classical physics can be viewed as a triumph of the idea that mind should be excluded from science, or at least from the physical sciences. Although the founders of modern science, such as Descartes and Newton, were not so rash as to proclaim that mind has nothing to do with the unfolding of nature, the scientists of succeeding centuries, emboldened by the spectacular success of the mechanical view of nature, were not so timid, and today we are seeing even in psychology a strong movement towards "materialism", i.e., toward the idea that "mind is brain". But while psychology has been moving toward the mechanical concepts of nineteenth-century physics, physics itself has moved in just the opposite direction.

The mentalistic bias of contemporary physics is perhaps best summarized in Heisenberg's statement that

> we are finally led to believe that the laws of nature that we formulate mathematically in quantum theory deal no longer with the particles themselves but with our knowledge of the elementary particles ... The conception of the objective reality of the particles has thus evaporated in a curious way, not into a fog of some new, obscure, or not yet understood reality concept, but into the transparent clarity of a mathematics that represents no longer the behavior of the elementary particles but rather our knowledge of this behaviour.[1]

This shift in the physicist's conception of nature, or at least in his conception of his *theory* about nature, away from the mechanical and toward the experiential, is expressed also by Bohr's statements:

> In our description of nature the purpose is not to disclose the real essence of phenomena but only to track down as far as possible relations between the multifold aspects of experience.[2]
>
> ... the goal of science is to augment and order our experience ...[3]

Bohr and Heisenberg each sought to deflate the idea that either he, or quantum theory itself, was asserting that the character of nature herself was essentially mental. Bohr emphasized that quantum theory was merely a tool for making predictions about our experiences:

> Strictly speaking, the mathematical formalism of quantum mechanics and electrodynamics merely offers rules of calculation for the deduction of expectations about observations obtained under well defined conditions specified by classical physical concepts.[4]

Heisenberg went even further:

> If we want to describe what happens in an atomic event we have to realize that the word "happens"... applies to the physical not the psychical act of observation, and we may say that the transition from the "possible" to the "actual" takes place as soon as the interaction between the [atomic] object and the measuring device, and thereby with the rest of the world, has come into play; it is not connected with the act of registration of the result in the mind of the observer. The discontinuous change in the probability function, however, takes place with the act of registration, because it is the discontinuous change in our knowledge in the instant of registration that has its image in the discontinuous change in the probability function.[5]

The final sentence affirms Heisenberg's position that the mathematical probability function of quantum theory represents "our knowledge". However, the statements that precede it affirm his belief that there are also some real "happenings" outside the minds of the human observers, and that these external events have the character of transitions of the "possible" to the "actual".

To describe these external events themselves in mathematical form one can introduce the idea of an *objective wave function* — a wave function that is like the one of Bohr and Heisenberg with respect to its mathematical properties (i.e., evolution via the Schrödinger equation etc.), but that represents the external world itself, and changes when the transitions from "possible" to "actual" take place, rather than with the registration of a result in the mind of the observer/scientist. This procedure would seem to be a reasonable step toward providing a conceivable description of nature herself, since it would allow the detailed and precise mathematical properties represented in quantum theory to be understood directly as mathematical characteristics of the world itself. This transformation can be termed the *ontologicalization* of quantum theory: it converts that theory from a structure conceived to be a mere tool for scientists — a tool to be used for very limited purposes — to a putative description of nature herself.

If we follow this tack, and endeavor to construe the mathematical structure represented by quantum theory as a feature of the world itself, then we may ask: What is the *nature* of that world? What *sort* of world do we live in?

The world represented by an ontogically interpreted quantum theory, with the quantum jumps representing transitions from "possible" to "actual", would be a strange sort of beast. The evolving quantum state, although controlled in part by mathematical laws that are direct analogs of the laws that in classical physics govern the motion of "matter", no longer represents anything substantive. Instead, this evolving quantum state would represent the "potentialities" and "probabilities" for actual events. Thus the "primal stuff" represented by the evolving quantum state would be idealike in character rather than matterlike, apart from its conformity to mathematical rules. On the other hand, mathematics has seemed, at least since the time of Plato, to be more a resident of a world of ideas, than a structure in the world of matter. Hence even this mathematical aspect of nature can be regarded as basically idealike. Indeed, quantum theory provides a detailed and explicit example of how an idealike primal stuff can be controlled in part by mathematical rules based in spacetime.

The actual events in quantum theory are likewise idealike: each such happening is a *choice* that selects as the "actual", in a way not controlled by any known or purported mechanical law, one of the potentialities generated by the quantum-mechanical law of evolution.

In view of these uniformly idealike characteristics of the quantum-physical world, the proper answer to our question "What sort of world do we live in?" would seem to be this: "We live in an *idealike* world, not a matterlike world." The material aspects are exhausted in certain mathematical properties, and these mathematical features can be understood just as well (and in fact better) as characteristics of an evolving idealike structure. There is, in fact, in the quantum universe no natural place for matter. This conclusion, curiously, is the exact reverse of the circumstance that in the classical physical universe there was no natural place for mind.

These remarks may appear to be nothing but a word game. But I think not. The change in our words indicates a change in our perception. By changing our perception of the kind of world we live in we change our perception of the possibilities. If some of the possibilities opened up by this altered perception of the basic nature of the physical world can be actualized within science then this change of words and perceptions will certainly count for something.

One possibility immediately opened up by this change is the possibility of integrating human consciousness into the physical sciences. This possibility was effectively blocked off when physical science meant, in the final analysis, classical physics. For there is an enormous conceptual gulf between the classical physicist's conceptualization of the physical world and the psychologist's conceptualization of the mental world. The essence of the classical physicist's conception of matter is its local-reductionistic nature: the idea the physical world can be decomposed into elementary local quantities that interact only with immediately adjacent neighbors. But conscious thoughts appear to be complex wholes, not merely at the functional level but also as directly experienced. Insofar as the experienced quality of a conscious thought constitutes its essence it is not possible to conceptualize a thought as a resident of the physical world, as that world was conceived of in classical physics. To bring a human conscious thought into the physicist's conception of the physical world one needs, within that conception, something having, *in its essence*, the integrity and complexity of that thought. The world as it is conceived of in classical physics is essentially reductive and therefore admits no essentially complex wholes.

This problem of unity is brought into clear focus by Daniel Dennett's book *Consciousness Explained.*[6] The thesis of the book is that brain processes proceed in "parallel pandemonium", with each of the processing units doing its own thing. The problem is then to bring the outputs of all these processes together into the integrated forms that we seem to experience in our stream of conscious thoughts. Dennett argues that this integration is, in fact, not possible, and hence that *our thoughts cannot be what they seem to be.*

This conclusion may indeed be what would emerge from a classical conception of what is going on in a human brain. But quantum theory opens completely new vistas. For the actual event in quantum theory can perfectly well be the actualization, *as a unit*, of an entire high-level pattern of neural firings. Such a pattern could have all of the complexity of a conscious thought, and yet be, in essence, a single actualized structure. From a logical point of view we have, therefore, the foundation of a rational way of linking conscious thoughts into the physicist's conception of nature.

It is, of course, one thing to have the logical basis of a rational way of integrating conscious thoughts into the physical sciences and another thing to have a consistent and coherent theory that really achieves this. There are the problems of explaining the linkage of brain states to the functional efficacy of the conscious thoughts and to the experiential qualities of conscious thoughts. Yet neither of these problems seems to be in principle beyond the bounds of rational explanation, within the quantum framework, which as explained earlier provides an intricate tapestry of idealike qualities.

The line of thinking described above has led to a serious attempt to bring human conscious experience into the quantum-mechanical description of nature.[7] This endeavor, though hardly complete, is, I believe, sufficiently successful to warrant considerable optimism as regards the prospects of ultimate success: a great deal of empirical information that had seemed very puzzling from a classical point of view now falls neatly into place.

In view of these developments I believe that the verdict of history will be that the Copenhagen interpretation was a half-way house: it was a right face that was the first step of an about face.

The scientific community has, rightly, a considerable amount of inertia. A complete turn around on the basic classical idea that mind should be rigorously excluded from the theory of the workings of the material universe was neither possible nor warranted during the 1920s and 1930s. Any attempt to correlate the revolutionary findings in the domain of atomic physics to the subtleties of the connection between mind and brain would have been extravagantly premature in view of the then-prevailing rudimentary state of our understanding of the workings of the brain. The appropriate course of action was first to see how far the new quantum ideas would carry us in domains that were under better empirical control.

During the past thirty years, however, the Copenhagen interpretation has lost a good deal of its hold on the minds of physicists. The words of Murray Gell-Mann give an indication of this shift:

> Niels Bohr brain-washed a whole generation of physicists into believing that the problem had been solved fifty years ago.[8]

The reasons for this change in attitude are many and diverse. One important reason is the expansion of the scope of quantum theory. The theory was originally designed to cover the domain of atomic physics, and was therefore concerned with things that were far beyond the range of our direct observation, and were thus approachable only indirectly with the aid of sophisticated measuring devices. Now, however, a problem that looms large in the minds of physicists is quantum gravity, which deals with quantum effects at the creation of the universe, and in the evolution of black holes. These phenomena are quite unlike the laboratory experiments in atomic physics that physicists were focussing on during the beginning of the century. The atomic-physics format of preparation-then-measurement fails to apply to these new problems. On the other hand, the ontological approach is far more demanding in terms of logical cohesion. The additional constraints imposed by the demand for a coherent ontology can provide guidance in our attempts to extend physical theory into the interesting new domains.

A second reason for the loosening of the grip of the Copenhagen interpretation is the fallout from the 1964 paper of John Bell.[9] The startling character of Bell's results caused physicists to take a careful look at the whole Bohr–Einstein controversy, and this left many of them with an uneasy sense that something important was perhaps being obscured by Bohr's subtle epistemological reasonings, which did not clearly do justice to the arguments of Einstein pertaining to locality.

A third reason for the fading influence of the Copenhagen interpretation is the construction by David Bohm of a thoroughly realistic model that reproduces all of the predictions of quantum theory.[10] This model laid to rest an opinion that was in the background of Copenhagen thinking, namely the idea that it was simply impossible to understand atomic phenomena in a realistic way. Although most physicists did not accept the idea that Bohm's simple model describes the way things really work, they were nonetheless quickly disabused of the impression that Bohr (or von Neumann) had showed that all realistic approaches were necessarily doomed to fail.

A fourth reason lies in the philosophical climate of the times. During the early part of the last century physicists were reeling from the impact of the loss of the "ether" and "absolute time". The whole idea that the universe could be understood in a completely clear mechanical way had been shattered. How could there be waves in a void: waves in a space devoid of medium? How could one understand the unfolding of our thoughts if there were no similar unfolding of nature herself; i.e., if the whole of spacetime history already "exists", as relativity theory seemed to require. The swallowing of such mysteries seemed to condition physicists not to balk at the even greater mysteries that quantum theory left unresolved. Furthermore, the parallel behavioristic movement in psychology, which also focussed on measurable quantities at the expense of any understanding of the unfolding stream of conscious thoughts, seemed to place all of science on the same operational track.

Now, however, the behaviorist approach to psychology seems to have failed, for technical reasons.[11] In psychology as in physics scientists are finding that increasingly complex models are needed to account for the complexity of the empirical data. But in the search for suitable complex models some orientation is needed. The data alone is insufficient: one needs some philosophy, and not merely an austere philosophy that recommends exclusive focussing upon the empirical facts obtained in a single narrow discipline. The insufficiency of the data in the various narrow disciplines, taken separately, is forcing scientists to bring into their theorizing information from an increasingly broad band of fields. Now in physics, for example, the problem of the innermost structure of the atoms is intertwined with the problem of the birth of the entire universe. Particle physics, astrophysics, and cosmology have merged into one field, at least at the level of theory.

Bold conceptions of large scope are needed to tie all these things together. The epistemological formulation of the Copenhagen interpretation seems, in the face of this complex situation, insufficiently helpful. Einstein's words in this connection are worth recalling:

It is my opinion that the contemporary quantum theory by means of certain definitely laid down basic concepts, which on the whole are taken over from classical physics, constitutes an optimum formulation of [certain] connections. I believe, however, that this theory offers no useful point of departure for future development.[12]

If what Einstein was judging to be insufficient was a science based upon the separation of the world into an ineffable nonclassical reality, and a then-unexplained classical character of our perceptions of that reality, then his judgement probably accords with the contemporary developments in science. But if, on the other hand, the nonclassical mathematical regularities identified by quantum theory are accepted as characteristics of the world itself, a world whose primal stuff is therefore essentially idealike, and if, moreover, these mathematical properties account in a natural and understandable way for the classical characteristic of our conscious perceptions, as they seem to do,[13] then we appear to have found in quantum theory the foundation for a science that may be able to deal successfully in a mathematically and logically coherent way with the full range of scientific thought, from atomic physics, to biology, to cosmology, including also the area that had been so mysterious within the framework of classical physics, namely the connection between processes in human brains and the stream of human conscious experience.

References

1 W. Heisenberg, *Daedalus* **87**, 95–108 (1958).
2 N. Bohr, *Atomic Physics and the Description of Nature* (Wiley, New York, 1934), p. 18.
3 N. Bohr, *Atomic Physics and Human Knowledge* (Wiley, New York, 1958), p. 60.
4 N. Bohr, *Essays 1958/62 on Atomic Physics and Human Knowledge* (Wiley, New York, 1963).
5 W. Heisenberg, *Physics and Philosophy* (Harper and Row, New York, 1958), p. 54.
6 D. C. Dennett, *Consciousness Explained* (Little, Brown and Co., New York, 1991).
7 H. P. Stapp, the present book.
8 M. Gell-Mann, in *The Nature of the Physical Universe, the 1976 Nobel Conference* (Wiley, New York, 1979), p. 29.
9 J. S. Bell, On the Einstein–Podolsky–Rosen Paradox, *Physics* **1**, 195 (1964).
10 D. Bohm, A Suggested Interpretation of Quantum Theory in Terms of "Hidden" Variables I, *Phys. Rev.* **85**, 166–179 (1952).
11 B. J. Baars, *The Cognitive Revolution in Psychology* (Guildford, New York, 1986).

12 A. Einstein, in *Albert Einstein Philosopher-Scientist*, edited by P. A. Schilpp (Tudor, New York, 1951).
13 H. P. Stapp, the present book.

Part IV

New Developments and Future Visions

11 Neuroscience, Atomic Physics, and the Human Person

This article is an integration of the contents of three talks and one text that I have prepared and delivered during the past year. They were aimed at four different audiences. The first talk was at a small conference in Philadelphia of scientists who are leading proponents of various diverse efforts to further develop and understand quantum theory. The second talk was at a public event in Switzerland where a number of scientists, and several artists, described to a general audience recent developments aimed at a better understanding of the nature of the human person. The third talk was at a conference in Tucson entitled "Quantum Approaches to the Understanding of Consciousness" and attended mainly by physicists, psychologists, and neuroscientists. The "text" was a section of a chapter of a book aimed at neuroscientists. Although the details of these four presentations were different, the essential content was the same: an explanation of the enormous difference in the scientific conception of the connection between mind and brain brought about by the replacement of the essentially seventeenth-century classical physical theory of Newton, Galileo, and Descartes by the twentieth-century quantum physics of Bohr, Heisenberg, Pauli, and von Neumann.

The orientations of the four presentations were varied. I began my talk in Switzerland with the words:

This talk is about you as a human person. It is about science's conception of you as a human person. It is about what makes you different from a machine. It is about your mind, and how your mind influences your bodily actions.

The talk in Philadelphia began with the words:

This talk has five closely related themes.
1 The most important development in science in the twenty-first century will be a deepening of our understanding of the nature of human beings.
2 The key unsolved question, there, is the nature of the connection between the mind and the brain.
3 Von Neumann's Processes I and II, applied to the human person, constitute genuine causal top–down and bottom–up mind–brain connections, respectively.
4 Process I involves "free choices".

5   These "free choices" can influence brain–body behavior.

The talk at Tucson began with:

Neuroscience is an important component of the scientific attack on the problem of consciousness. However, most neuroscientists, viewing our discussions, see only dissent and discord, and no reason to believe that quantum theory has any profound relevance to the dynamics of the conscious brain. It is therefore worthwhile, in this first plenary talk of the 2003 Tucson conference on "Quantum Approaches to the Understanding of Consciousness", to focus on the central issue, which is the crucial role of "the observer", and more specifically, "the mind of the observer", in contemporary physical theory. I shall therefore review this radical departure of present-day basic physics from the principles of classical physics, and then spell out some of its ramifications for neuroscience.

The section of the chapter of the book aimed at neuroscientists was part of a chapter describing recent experiments involving the conscious control of emotions, and the large differences in brain activity when a conscious effort is made — or is not made — to suppress the emotional impact of certain visual stimuli. The experiments show strong correlations between data of two distinct kinds: (1) recordings on devices that are measuring physical properties of the brain of a subject, and (2) instructions to those subjects couched in psychological terms pertaining to mental efforts and strategies. The section explains the new modes of understanding and modeling the correlations between data of these two disparate kinds created by the orthodox (von Neumann) quantum theoretic conceptualization of the conscious brain, as contrasted to the classical conceptualization. That section stresses the close similarity between the situations faced by atomic scientists and neuroscientists in their attempts to understand in causal terms the correlations between data described in psychological and physical terms, and how quantum theory provides for bona fide top–down influences of mental actions upon neural processes, and also an operationally and pragmatically simpler theory of the conscious brain that both rests upon and emerges from contemporary physics.

The present article is aimed at all of those audiences, and addresses all of those topics.

I have had to include a few key equations, in order to allow physicists to know exactly what I am saying, but have described in ordinary words what these equations mean. I believe that these symbolic expressions will be helpful to all readers, even those who proclaim deep-seated eternal aversion to math.

Before proceeding I should indicate what I mean by the words "mind" and "brain".

Your *mind* is your stream of consciousness. It consists of your thoughts, ideas, and feelings, and is described in *psychological* or *mental* terms.

Your *brain* is an organ in your body consisting of nerve cells and other tissues, and is described in *physical* terms — in terms of *properties assigned to tiny spacetime regions inside your skull.*

Your mind and your brain are obviously related. Your conscious thought can cause your arm to rise. What happens is this: Your conscious intentional effort causes nerve pulses to emanate from your brain, and these pulses cause muscles in your arm to contract, and those contractions cause your arm to rise.

But *how*, according to the basic principles of science, does your conscious thought initiate that chain of bodily events? How does a *mental* action cause *physical* events?

The central theme of all four presentations, and of this article, is the tremendous difference in the scientific understanding of the dynamics of the conscious brain that emerges from orthodox quantum theory, with its essential introduction of the active human agent-participant, as contrasted to classical physics. Although many neuroscientists and neurophilosophers do not explicitly specify that they are assuming the validity of classical physics, which they know to be false in the regime of the behaviors of the ions and molecules that play a key role in the dynamics of the conscious brain, they nevertheless endeavor to conceptualize the dynamics of the conscious brain in essentially classical terms: they have closed their minds to the huge practical and conceptual advantages wrought by the twentieth-century advances in physics. To reveal what they are losing it is helpful first to review the precepts of classical physics.

11.1 Classical Physics

Classical physics is a theory of nature that originated with the work of Isaac Newton in the seventeenth century and was advanced by the contributions of James Clerk Maxwell and Albert Einstein. Newton based his theory on the work of Johannes Kepler, who found that the planets appeared to move in accordance with a simple mathematical law, and in ways wholly determined by their spatial relationships to other objects. *Those motions were apparently independent of our human observations of them.*

Newton assumed that all physical objects were made of tiny miniaturized versions of the planets, which, like the planets, moved in accordance with simple mathematical laws, independently of whether we were aware of them or not. He found that he could explain the motions of the planets, and also the motions of large terrestrial objects and systems, such as cannon balls, falling apples, and the tides, by assuming that every tiny planetlike particle in the solar system attracted every other one with a force inversely proportional to the square of the distance between them.

This force was *an instantaneous action at a distance*: it acted instantaneously, no matter how far apart the particles were located. This feature troubled Newton. He wrote to a friend:

> That one body should act upon another through the vacuum, without the mediation of anything else, by and through which their action and force may be conveyed from one to another, is to me so great an absurdity that I believe no man, who has in philosophical matters a competent faculty of thinking, can ever fall into it.[1]

Although Newton's philosophical persuasion on this point is clear, he nevertheless formulated his universal law of gravity without specifying how it was mediated.

Albert Einstein, building on the ideas of Maxwell, discovered a suitable mediating agent: a distortion of the structure of spacetime itself. Einstein's contributions made classical physics into what is called a *local theory*: there is no action at a distance. All influences are transmitted essentially by contact interactions between tiny neighboring mathematically described "entities", and no influence propagates faster than the speed of light.

Classical physics is, moreover, *deterministic*: the interactions are such that the state of the physical world at any time is completely determined by the state at any earlier time. Consequently, according to classical theory, the complete history of the physical world *for all time* is mechanically fixed by contact interactions between tiny component parts, together with the initial condition of the primordial universe.

This result means that, according to classical physics, *you are a mechanical automaton*: your every physical action was pre-determined before you were born solely by mechanical interactions between tiny mindless entities. Your mental aspects are *causally redundant*: everything you do is completely determined by mechanical conditions alone, without reference to your thoughts, ideas, feelings, or intentions. Your intuitive feeling that your mental intentions make a difference in what you do is, according to the principles of classical physics, a false and misleading illusion.

Many scientists, philosophers, writers, intellectuals, teachers, and policy makers claim to believe this mechanical conception of human beings, and base policies upon it. They believe that this is what science says, and hence that this is what you must believe. *But this is not what science says!* It is what *classical physics* says! It is what an essentially seventeenth-century precursor to contemporary physical theory says!

There are two ways within classical physics to understand this total incapacity of your mental side—your stream of consciousness—to make any difference in what you do. The first is to consider your thoughts, ideas, and feelings to be epiphenomenal *by-products* of the activity of your brain. Your mental side is then a causally impotent sideshow that is *produced*, or *caused*, by your brain, but that generates no reciprocal action back upon your brain. The second way is to contend that your mental aspects are the *very same things* as certain kinds of motions of various tiny parts of your brain.

11.2 Problems with the Classical Physics Idea of the Conscious Brain

William James reasoned against the first possibility, epiphenomenal consciousness, by arguing that

> *The particulars of the distribution of consciousness*, so far as we know them, *point to its being efficacious.*[2]

He noted that consciousness seems to be

> an organ, superadded to the other organs which maintain the animal in its struggle for existence; and the presumption of course is that it helps him in some way in this struggle, just as they do. But it cannot help him without being in some way efficacious and influencing the course of his bodily history.

James said that the study described in his book

> will show us that consciousness is at all times primarily a *selecting agency*.

It is present when choices must be made between different possible courses of action. He further mentioned that

> It is to my mind quite inconceivable that consciousness should have *nothing to do* with a business to which it so faithfully attends.[3]

If consciousness has no effect upon the physical world, then what keeps a person's mental world aligned with his physical situation: what keeps his pleasures in general alignment with actions that benefit him, and pains in general correspondence with things that damage him, if pleasure and pain have no effect at all upon his actions?

These liabilities of the notion of epiphenomenal consciousness lead many thinkers to turn to the alternative possibility that a person's stream of consciousness is the *very same thing* as some activity in his brain: consciousness is an "emergent property" of brains.

A huge philosophical literature has developed arguing for and against this idea. The primary argument against this "emergent-identity theory" position, *within a classical physics framework*, is that within classical physics the full description of nature is in terms of numbers assigned to tiny space-time regions, and there appears to be no way to understand or explain how to get from such a restricted conceptual structure, which involves such a small part of the world of experience, to the whole. How and why should that extremely limited conceptual structure, which arose basically from idealizing, by miniaturization, certain features of observed planetary motions—and which is now known to be profoundly incorrect in physics—suffice to explain the totality of experience, with its pains, sorrows, hopes, colors, smells, and moral judgments? Why, given the known failure of classical physics at the fundamental level, should that richly endowed whole be explainable in terms of such a narrowly restricted part?

The core ideas of the arguments in favor of an identity-emergent theory of consciousness are illustrated by Roger Sperry's example of a "wheel".[4] A wheel obviously does something: it is causally efficacious; it carries the cart. It is also an *emergent property*: there is no mention of "wheelness" in the formulation of the laws of physics, and "wheelness" did not exist in the early universe; "wheelness" *emerges* only under certain special conditions. And the macroscopic wheel exercises "top–down" control of its tiny parts. All these properties are perfectly in line with classical physics, and with the idea that "a wheel is, precisely, a structure constructed out of its tiny atomic parts." So why not suppose "consciousness" to be, like "wheelness", an emergent property of its classically conceived tiny physical parts?

The reason that consciousness is not analogous to wheelness, *within the context of classical physics*, is that the properties that characterize wheelness are properties that are *entailed*, within the conceptual framework of classical physics, by properties specified in classical physics, whereas the properties that characterize consciousness, namely the way it feels, are not entailed, within the conceptual structure provided by classical physics, by the properties specified by classical physics.

This is the huge difference-in-principle that distinguishes consciousness from things that, according to the precepts of classical physics, are constructible out of the particles that are postulated to exist by classical physics.

Given the state of motion of each of the tiny physical parts of a wheel, as it is conceived of in classical physics, the properties that characterize the wheel—e.g. its roundness, radius, center point, rate of rotation, etc.— are specified within the conceptual framework provided by the principles of classical physics, *which specify only geometric-type properties such as changing locations and shapes of conglomerations of particles, and numbers assigned to points in space.* But given the state of motion of each tiny part of the brain, as it is conceived of in classical physics, the properties that characterize a stream of consciousness—the painfulness of the pain, the feeling of the anguish, or of the sorrow, or of the joy—are not specified, within the conceptual framework provided by the principles of classical physics. Thus it is possible, within that classical physics framework, to strip away those feelings without disturbing the physical descriptions of the motions of the tiny parts. One can, *within the conceptual framework of classical physics*, take away the consciousness without affecting the locations and motions of the tiny physical parts of the brain. But one cannot, within the conceptual framework provided by classical physics, take away the wheelness of the wheel without affecting the locations and motions of the tiny physical parts of a wheel.

Because one can, within the conceptual framework provided by classical physics, strip away the consciousness without affecting the physical behavior, one cannot rationally claim that the consciousness is the cause of the physical behavior, or is *causally efficacious* in the physical world. Thus the "identity theory" or "emergent property" strategy fails in its attempt to make consciousness efficacious, within the conceptual framework provided by classical physics. Moreover, the whole endeavor to base brain theory on classical physics is undermined by the fact that the classical theory fails to work for phenomena that depend critically upon the properties of the atomic constituents of the behaving system, and brains are such systems: brain processes depend critically upon synaptic processes, which depend critically upon ionic processes that are highly dependent upon their quantum nature. This essential involvement of quantum effects will be discussed in detail in later sections.

11.3 The Quantum Approach

Classical physics is an *approximation* to a more accurate theory—called quantum mechanics—and quantum mechanics makes mind efficacious. Quantum mechanics *explains* the causal effects of mental intentions upon physical systems: it *explains* how your mental effort can produce the brain events that cause your bodily actions. Thus quantum theory converts science's picture of you from that of a mechanical automaton to that of a mindful human person. Quantum theory also shows, explicitly, how the approximation that reduces quantum theory to classical physics completely eliminates all effects of your conscious thoughts upon your brain and body.

Hence, from a physics point of view, trying to understand the mind–brain connection by going to the classical approximation is absurd: it amounts to trying to understand something in an approximation that eliminates the effect you are trying to study.

Quantum mechanics arose during the twentieth century. Scientists discovered, empirically, that the principles of classical physics were not correct. Moreover, they were wrong in ways that no minor tinkering could ever fix. The *basic principles* of classical physics were thus replaced by *new basic principles* that account uniformly both for all the successes of the older classical theory and also for all the newer data that is incompatible with the classical principles.

11.3.1 Physical Theory Was Turned Inside Out

The most profound alteration of the fundamental principles was to bring the consciousness of human beings into the basic structure of the physical theory. In fact, the whole *conception of what science is* was turned inside out. The core idea of classical physics was to describe the "world out there", with no reference to "our thoughts in here". But the core idea of quantum mechanics is to describe *our activities as knowledge-seeking and knowledge-using agents*. Thus quantum theory involves, basically, not just what is "out there", but also what is "in here", namely "our knowledge". Consciousness is thus introduced into contemporary orthodox physical theory, not as something *whose existence needs to be explained*, but rather as something whose detailed structure and detailed connection to brain activities needs to be further explicated.

11.3.2 Science Must Bridge the Psychophysical Divide

The basic philosophical shift in quantum theory is the *explicit* recognition that science is about *what we can know*. It is fine to have a beautiful and elegant mathematical theory about an imagined *"really existing physical world out there"* that meets a lot of intellectually satisfying criteria. But the essential demand of science is that the theoretical constructs be tied to the experiences of the human scientists who devise ways of testing the theory, and of the human engineers and technicians who both participate in these tests and eventually put the theory to work. So the structure of a proper physical theory must involve not only the part describing the behavior of the not-directly-experienced theoretically postulated entities, expressed in some appropriate symbolic language, but also a part describing the human experiences that are involved in these tests and applications, expressed in the language that we actually use to describe such experiences to ourselves and each other. Finally we need some "bridge laws" that specify the connection between the concepts described in these two different languages.

Classical physics met these requirements in a rather trivial kind of way, with the relevant experiences of the human participants being taken to be direct apprehensions of various gross behaviors of large-scale properties of big objects composed of huge numbers of the tiny atomic-scale parts. And these apprehensions were taken to be passive: they had no effect on the behaviors of the systems being studied. But the physicists who were examining the behaviors of systems that depend sensitively upon the behaviors of their tiny atomic-scale components found themselves forced to go to a less trivial theoretical arrangement, in which the human agents were no longer passive observers but were *active participants* in ways that contradicted, and were impossible to comprehend within, the general framework of classical physics, *even when the only features of the physically described world that the human beings observed were large-scale properties of measuring devices.*

11.3.3 The Two-Way Quantum Psychophysical Bridge

The sensitivity of the behavior of the devices to the behavior of some tiny atomic-scale particles propagates in such a way that the acts of observation by the human observers of *large-scale properties of the devices* could no longer be regarded as passive: these acts were assigned a crucial selective action. Thus the core structure of the basic general physical theory became transformed in a profound way: the connection between physical behavior and human knowledge was changed from a one-way bridge to a mathematically specified two-way interaction that involves *selections* performed by conscious minds.

This profound change in the principles is encapsulated in Niels Bohr's dictum that

in the great drama of existence we ourselves are both actors and spectators.[5]

The emphasis here is on "actors": in classical physics we, and in particular our minds, were mere spectators.

This revision must be expected to have important ramifications in neuroscience, because the issue of the connection between mind (the psychologically described aspects of a human being) and brain/body (the physically described aspects of that person) has recently become a matter of central concern in neuroscience.

11.4 The Copenhagen Formulation

The original formulation of quantum theory was created mainly at an Institute in Copenhagen directed by Niels Bohr and is called "The Copenhagen Interpretation". Owing to the profound strangeness of the conception of nature entailed by the new mathematics, the Copenhagen strategy was to refrain from making ordinary ontological claims, but to take, instead, a fundamentally pragmatic stance. Thus the theory was formulated *basically* as a set of practical rules for how scientists should go about their tasks of acquiring knowledge, and then using this knowledge in practical ways. Speculations about "what the world out there—apart from our knowledge of it—is really like" were regarded as "metaphysics", and hence outside real science.

Copenhagen quantum theory is about the relationships between human agents (called "participants" by John Wheeler) and the systems that they act upon. In order to achieve this conceptualization the Copenhagen formulation separates the physical universe into two parts, which are described in two different languages. One part is the observing human agent and his measuring devices. That part is described in mental terms—in terms of our instructions to colleagues about how to set up the devices, and our reports of what we then learn. The other part of nature is *the system that the agent is acting upon*. That part is described in physical terms—in terms of mathematical properties assigned to tiny spacetime regions.

11.4.1 Von Neumann's Process II

The great mathematician and logician John von Neumann formulated Copenhagen quantum theory in a rigorous way.

Von Neumann identified two very different processes that enter into the quantum theoretical description of the evolution of a physical system. He called them Process I and Process II.[6] Process II is the analog in quantum theory of the process in classical physics that takes the state of a system at one time to its state at a later time. This Process II, like its classical analog, is *local* and *deterministic*. However, Process II by itself is not the whole story: it generates physical worlds that do not agree with human experiences. For example, if Process II were the *only* process in nature, then the quantum state of the moon would represent a structure smeared out over a large part of the sky.

11.4.2 Process I: A Dynamical Psychophysical Bridge

To tie the quantum mathematics to human experience in a rationally coherent and mathematically specified way quantum theory introduces *another process*, which von Neumann calls Process I. It is a *selection* process that is tied to conscious experience, and it is not determined by the micro-local deterministic Process II. It is a selection made by an agent about how he or she will act or attend.

Any physical theory must, in order to be complete, specify how the elements of the theory are connected to human experience. In classical physics this connection is part of a *metaphysical* superstructure: it is not part of the core dynamical description. But in quantum theory this connection of the mathematically described physical state to conscious experiences is part of the essential dynamical structure. And this connecting process is not passive: it does not represent a mere *witnessing* of a physical feature of nature by a passive mind. Rather, the process is active: it injects into the physical state of the system being acted upon properties that depend upon the intentional chosen action of the observing agent.

Quantum theory is built upon the practical concept of intentional actions by agents. Each such action is expected or intended to produce an experiential response or feedback. For example, a scientist might act to place a Geiger counter near a radioactive source, and expect to see the counter either "fire" during a certain time interval or not "fire" during that interval. The experienced response, "Yes" or "No", to the question "Does the counter fire during the specified interval?" specifies one bit of information. Quantum theory is thus an information-based theory built upon the knowledge-acquiring actions of agents, and the knowledge that these agents thereby acquire.

Probing actions of this kind are performed not only by scientists. Every healthy and alert infant is engaged in making willful efforts that produce experiential feedbacks, and he or she soon begins to form expectations about what sorts of feedbacks are likely to follow from some particular kind of effort. Thus both empirical science and normal human life are based on paired realities of this action–response kind, and our physical and psychological theories are both basically attempts to understand these linked realities within a rational conceptual framework.

The basic building blocks of quantum theory are, then, a set of intentional actions by agents, and for each such action an associated collection of possible "Yes" feedbacks, which are the possible responses that the agent can judge to be in conformity to the criteria associated with that intentional act. For example, the agent is assumed to be able to make the judgment

"Yes" the Geiger counter clicked or "No" the Geiger counter did not click. And he must be able to report "Yes" the counter is in the specified place, or "No" it is not there. Science would be difficult to pursue if scientists could make no such judgments about what they were experiencing.

All known physical theories involve idealizations of one kind or another. In quantum theory the main idealization is not that every object is made up of miniature planetlike objects. It is rather that there are agents that perform intentional acts each of which can result in a feedback that may conform to a certain criterion associated with that act. One bit of information is introduced into the world in which that agent lives, according to whether the feedback conforms or does not conform to that criterion. Thus knowing whether the counter clicked or not places the agent on one or the other of two alternative possible separate branches of the course of world history.

These remarks reveal the enormous difference between classical physics and quantum physics. In classical physics the elemental ingredients are tiny invisible bits of matter that are idealized miniaturized versions of the planets that we see in the heavens, and that move in ways unaffected by our consciousness, whereas in quantum physics the elemental ingredients are intentional actions by agents, the feedbacks arising from these actions, and the effects of our actions on the physical systems that our actions act upon.

Consideration of the character of these differences makes it plausible that quantum theory may be able to provide the foundation of a scientific theory of the human person that is better able than classical physics to integrate the physical and psychological aspects of his nature. For quantum theory describes the effects of a person's intentional actions upon the physical world, whereas classical physics systematically leaves these effects out.

An intentional action by a human agent is partly an intention, described in psychological terms, and partly a physical action, described in physical terms. The feedback also is partly psychological and partly physical. In quantum theory these diverse aspects are all represented by logically connected elements in the mathematical structure that emerged from the seminal discovery of Heisenberg. That discovery was that in order to get the quantum generalization of a classical theory one must formulate the theory in terms of actions. A key difference between *numbers* and *actions* is that if A and B are two actions, then AB represents the action obtained by performing the action A upon the action B. If A and B are actions, then, generally, AB is different from BA: the order in which actions are performed matters.

The intentional actions of agents are represented mathematically in Heisenberg's space of actions. Here is how it works.

Each intentional action depends, of course, on the intention of the agent, and upon the state of the system upon which this action acts. Each of these two aspects of nature is represented within Heisenberg's space of actions by an action.

The idea that a "state" should be represented by an "action" may sound odd, but Heisenberg's key idea was to replace what classical physics took to be a "being" by a "doing". I shall denote the action that represents the state being acted upon by the symbol S.

An intentional act is an action that is intended to produce a feedback of a certain conceived or imagined kind. Of course, no intentional act is sure-fire: one's intentions may not be fulfilled. Hence the intentional action puts in play a process that will lead either to a confirmatory feedback "Yes", the intention is realized, or to the result "No", the "Yes" response failed to occur.

The effect of this intentional mental act is represented mathematically by an equation that is one of the key equations of quantum theory. This equation represents, within the quantum mathematics, the effect of the Process I mental action upon the quantum state S of the system being acted upon. The equation is

$$S \to S' = PSP + (1 - P)S(1 - P).$$

This formula exhibits the important fact that this Process I action changes the state S of the system being acted upon into a new state S', which is a sum of two parts.

The first part, PSP, represents the possibility in which the experiential feedback called "Yes" appears, and the second part, $(1 - P)S(1 - P)$, represents the alternative possibility "No", this feedback does not appear. Thus the intention of the action and the associated experiential feedback are tied into the mathematics that describes the dynamics of the physical system being acted upon.

The action P is important. It represents an action upon the system that is being acted upon by the agent, and it depends on the *intention of the agent*. The action represented by the symbol P, acting both on the right and on the left of S, is the action of eliminating from the state S all parts of S except the "Yes" part. That particular retained part is determined by the intentional choice of the agent. The action of $(1 - P)$, acting both on the right and on the left of S, is, analogously, to eliminate from S all parts of S except the "No" parts.

The projection operator P is required to satisfy $P = PP$. This implies that $P(1 - P) = (1 - P)P = 0$, which says that the sequence of these two actions, P and $(1 - P)$, in either order, leave nothing.

Thus the action P is an action in the space in which the physical system is represented, and it reduces to zero all components that correspond to the "No" response, but leaves intact the components corresponding to the "Yes" response to the intentional action. The action of $(1 - P)$ is the analogous action with "Yes" and "No" interchanged. The action of P is the representation of an intentional mental action upon a physically described system.

Notice that Process I produces the *sum* of the two alternative possible feedbacks, not just one or the other. Since the feedback must either be "Yes" or "No = Not-Yes", one might think that Process I, which *keeps* both the "Yes" and the "No" parts, would do nothing. But that is not correct! This is a key point. It can be verified by noticing that S can be written as a sum of four parts, only two of which survive the Process I action:

$$S = PSP + (1 - P)S(1 - P) + PS(1 - P) + (1 - P)SP.$$

This formula is a strict identity. The dedicated reader can easily confirm it by collecting the contributions of the four occurring terms PSP, PS, SP, and S, and verifying that all terms but S cancel out. This identity shows that the state S can be expressed as a sum of four parts, *two of which are eliminated by Process I.*

But this means that Process I has a *nontrivial* effect upon the state being acted upon: it eliminates the two terms that correspond neither to the appearance of a "Yes" feedback nor to the failure of the "Yes" feedback to appear.

That is the first key point: quantum theory has a specific dynamical process, Process I, which specifies the effect upon a physically described system of an *intentional act* by a conscious agent.

11.4.3 Free Choices

The second key point is this: the agent's choices are "free choices", *in the specific sense specified below*.

Orthodox quantum theory is formulated in a realistic and practical way. It is structured around the activities of human agents, who are considered able to freely elect to probe nature in any one of many possible ways. Bohr emphasized the freedom of the experimenters in passages such as:

> The freedom of experimentation, presupposed in classical physics, is of course retained and corresponds to the free choice of experimental arrangement for which the mathematical structure of the quantum mechanical formalism offers the appropriate latitude.[7]

This freedom of action stems from the fact that in the original Copenhagen formulation of quantum theory the human experimenter is considered to stand outside the system to which the quantum laws are applied. Those quantum laws are the only precise laws of nature recognized by that theory. Thus, according to the Copenhagen philosophy, *there are no presently known laws that govern the choices* made by the agent/experimenter/observer/participant about how the observed system is to be probed. This choice is, *in this very specific sense*, a "free choice". It is not ruled out that some deeper theory will eventually provide a causal explanation of this "choice".

11.4.4 Probabilities

The predictions of quantum theory are generally statistical: only the *probabilities* that the agent will experience each of the alternative possible feedbacks are specified. Which of these alternative possible feedbacks will actually occur in response to a Process I action is not determined by quantum theory.

The formula for the probability that the agent will experience the feedback "Yes" is

$$\mathrm{tr}\, PSP/\mathrm{tr}\, S$$

where the symbol tr represents the trace operation. This trace operation means that the actions act in a cyclic fashion, so that the rightmost action acts back around upon the leftmost action. Thus, for example,

$$\mathrm{tr}\, ABC = \mathrm{tr}\, CAB = \mathrm{tr}\, BCA.$$

The product ABC represents the result of letting A act upon B, and then letting that product AB act upon C. But what does C act upon? Taking the trace of ABC means specifying that C acts back around on A.

An important property of a trace is that the trace of any of the sequences of actions that we consider must always give a positive number or zero. Thus this trace operation is what ties the actions, as represented in the mathematics, to measurable numbers.

(The trace operation, and in fact the operation of multiplying together any two operators, is the quantum analog of the classical process of integrating over all of "phase space", giving equal *a priori* weighting to equal volumes of phase space. Thus the trace operation is in effect a statistical sum over all of the "loose ends" that are not fixed in the expression upon which the trace operation acts.)

11.5 Von Neumann's Psychophysical Theory of the Conscious Brain

The Copenhagen approach separates the world into two parts: "the Observer", which includes the mind, brain, and body of the personal observer together with his measuring devices; and "the System" that this observer is acting upon. "The Observer" is described in psychological terms, whereas "the System" is described in physical/mathematical spacetime terms.

This procedure works very well in practice. However, it seems apparent that the body and brain of the human agent, and his devices, are parts of the physical universe. Hence a complete theory ought to be able to include our bodies and brains in the physically described part of the theory. On the other hand, the structure of the theory depends critically also upon the features that are represented in Process I, and that are described in mentalistic language as intentional actions and experiential feedbacks.

Von Neumann showed that it was possible, without significantly disturbing the predictions of the theory, to shift the bodies and brains of the agents, along with their measuring devices, into the physical world, *while retaining, and ascribing to the mind of the agent, those mentalistically described properties of the agents that are essential to the structure of the theory.* The system acted upon by the mind is the brain. Thus in this von Neumann re-formulation the Process I action is an action of mind upon brain. Hence von Neumann's re-formulation provides us with the core of a science-based dynamical theory of the conscious brain.

It is worthwhile to reflect for a moment on the ontological aspects of von Neumann quantum theory. Von Neumann himself, being a clear thinking mathematician, said very little about ontology. But he called the mentalistically described aspect of the agent his "abstract 'ego' ".[8] This phrasing tends to conjure up the idea of a disembodied entity, standing somehow apart from the body/brain. But another possibility is that consciousness is an *emergent property* of the body/brain. Notice that some of the problems that occur in trying to defend the idea of emergence within the framework of classical physical theory disappear when one accepts the validity of quantum theory. For one thing, one no longer has to defend against the charge that the emergent property, consciousness, has no "genuine" causal efficacy, because anything it does is done already by the physically described process, independently of whether the psychologically described aspect emerges or not. In quantum theory the causal efficacy of our thoughts is no illusion: it's the real thing!

Another difficulty with "emergence" in a classical physics context is in understanding how the motion of a set of miniature planetlike objects, careening through space, can *be a painful experience*. But within the quantum framework the basic physical structure, namely the quantum state, is essentially knowledge or information imbedded in spacetime. Hence there is no intrinsic problem with the idea that a sudden increment in a person's knowledge should be represented by a sudden jump in the quantum state of his brain. The identification of conscious actions with physical actions is no longer problematic. This is because the old idea of "matter" has been eradicated, and replaced by a mathematical representation of an information-based psychophysical reality.

In this connection, Heisenberg remarked:

> The conception of the objective reality of the elementary particles has thus evaporated not into the cloud of some obscure new reality concept, but into the transparent clarity of a mathematics that represents no longer the behavior of the particle but rather our knowledge of this behavior.[9]

11.5.1 Conservation of Causality

The question arises: How can the effect of a psychologically described action be injected into the dynamics of a physically described system without upsetting the causal structure of the latter.

The answer is this: Physicists have discovered an important and unexpected property of nature. It pertains to observable phenomena that depend upon microscopic properties that are *in principle inaccessible to observation*. In such a situation we are *in principle* unable, owing to the lack of crucial micro-data, to give a complete causal description of the observable phenomena. However, our principled inability to give a complete causal account of the psychologically described phenomena, owing to this inherent gap in the micro-data, can be partially offset by introducing into the theory, *instead of the inaccessible micro-data*, the *psychologically described selection of an action* made upon the system by an agent.

Thus the loss of causal determination at the microlevel, owing to the limitations imposed by Heisenberg's uncertainty principle, allows an alternative (statistical) causal account to be achieved by replacing the inaccessible micro-data by empirically available and controllable data about human selections of actions!

This feature discovered in atomic science should be equally important in neuroscience. That is because the basic problem in neuroscience is essentially the same as the one in atomic physics. In both cases the problem is to provide a causal account of connections between experiences that depend sensitively upon micro-properties that are in principle inaccessible. But quantum theory shows how the principled loss of information at the microlevel can be partially offset by using, instead, the controllable and reportable variables of the intentional actions of human beings. Nature left open a causal gap for us to occupy.

11.6 The Quantum Brain

The quantum state of a human brain is, of course, a very complex thing. But its main features can be understood by considering first a classical conception of the brain, and then folding in some key features that arise already in the case of the quantum state of a *single* particle, or object, or degree of freedom.

11.6.1 States of a Simple Harmonic Oscillator

One of the most important examples of a quantum state is the one corresponding to a pendulum, or more precisely, to what is called a "simple harmonic oscillator". Such a system is one in which there is a restoring force that tends to push the center of the object to a single "base point" of lowest energy, and in which the strength of this restoring force is directly proportional to the distance of the center point of the object from this base point.

According to classical physics any such system has a state of lowest energy. In this state the center point of the object lies motionless at the base point. In quantum theory this system again has a state of lowest energy, but the center point is not localized at the base point: it is represented by a *cloudlike* spatial structure that is spread out over a region that extends to infinity. However, the amplitude of this cloudlike form has the shape of a bell: it is largest at the base point, and falls off in a prescribed manner as the distance of the center point from the base point increases.

If one were to squeeze this state of lowest energy into a narrower space, and then let it loose, the cloudlike form would first explode outward, but then settle into an oscillating motion. Thus the cloudlike spatial structure behaves rather like a swarm of bees, such that the more they are squeezed in space the faster they move, and the faster the squeezed cloud will explode outward when the squeezing constraint is released. These visualizable properties extend in a natural way to many-particle cases.

11.6.2 The Double-Slit Experiment

An important difference between the behavior of the quantum cloudlike form and the somewhat analogous *classical probability distribution* is exhibited by the famous *double-slit experiment*. If one shoots an electron, an ion, or any other quantum counterpart of a tiny classical object, at a narrow slit, then if the object passes through the slit, the associated cloudlike form will fan out over a wide angle. But if one opens two closely neighboring narrow slits, then what passes through the slits is described by a probability distribution that is not just the sum of the two separate fanlike structures that would be present if each slit were opened separately. Instead, at some points the probability value will be *twice the sum* of the values associated with the two individual slits, and in other places the probability value drops nearly to zero, even though both individual fanlike structures give a large probability value at that place. These *interference* features of the quantum cloudlike structure make that structure logically different from a classical-physics probability distribution, for in the classical case the probabilities arising from the two slits would simply add, because, according to classical principles, the particle must pass through one slit or the other, and the fact that some other slit is also open should not matter very much.

Quantum theory deals consistently with this interference effect, and all the other nonclassical properties of these cloudlike structures.

11.7 Nerve Terminals, Ion Channels, and the Need to Use Quantum Theory

Some neuroscientists who study the relationship of consciousness to brain process believe that classical physics will be adequate for that task. That belief would have been reasonable during the nineteenth century, but now, in the twenty-first, it is rationally untenable: quantum theory must in principle be used because the behavior of the brain depends sensitively upon ionic and atomic processes, and these processes involve quantum effects.

To study quantum effects in brains within an orthodox (i.e., Copenhagen or von Neumann) quantum theory one must use the von Neumann formulation. The reason is that *Copenhagen* quantum theory is formulated in a way that leaves out the quantum dynamics of the human observer's body and brain. But von Neumann quantum theory takes the physical system S upon which the crucial Process I acts to be the brain of the agent, or some part of the brain. Thus Process I then describes an interaction between a person's stream of consciousness, described in mentalistic terms, and the activity in his brain, described in physical terms. That interaction drops completely out when one passes to the classical approximation. Hence ignoring quantum effects in the study of the mind–brain connection means, according to the basic principles of physics, ignoring the dynamical connection one is trying to study.

One must *in principle* use quantum theory. But there is then the quantitative issue of how important the quantum effects are.

To explore that question we now consider the quantum dynamics of nerve terminals.

11.7.1 Nerve Terminals and Ion Channels

Nerve terminals are essential connecting links between nerve cells. The way they work is quite well understood. When an action potential traveling along a nerve fiber reaches a nerve terminal, a host of ion channels open. Calcium ions enter through these channels into the interior of the terminal. These ions migrate from the channel exits to release sites on vesicles containing neurotransmitter molecules. The triggering effect of the calcium ions causes these contents to be dumped into the synaptic cleft that separates this terminal from a neighboring neuron, and these neurotransmitter molecules influence the tendencies of that neighboring neuron to "fire".

The channels through which the calcium ions enter the nerve terminal are called "ion channels". At their narrowest points they are less than a nanometer in diameter.[10] This extreme smallness of the opening in the ion channels has profound quantum mechanical importance. The consequence is essentially the same as the consequence of the squeezing of the state of the simple harmonic operator, or of the narrowness of the slits in the double-slit experiments. The narrowness of the channel restricts the lateral spatial dimension. Consequently, the lateral velocity is forced by the *quantum uncertainty principle* to become large. This causes the cloud associated with the calcium ion to *fan out* over an increasing area as it moves away from the tiny channel to the target region where the ion will be absorbed as a whole, or not absorbed, on some small triggering site.

This spreading of the ion wave packet means that the ion may or may not be absorbed on the small triggering site. Accordingly, the vesicle may or may not release its contents. Consequently, the quantum state of the vesicle has a part in which the neurotransmitter is released and a part in which the neurotransmitter is not released. This quantum splitting occurs at every one of the trillions of nerve terminals.

What is the effect of this *necessary* incursion of the cloudlike quantum character of the ions into the evolving state of the brain?

A principal function of the brain is to receive clues from the environment, to form an appropriate plan of action, and to direct and monitor the activities of the brain and body specified by the selected plan of action. The exact details of the plan will, for a classical model, obviously depend upon the exact values of many noisy and uncontrolled variables. In cases close to a bifurcation point the dynamical effects of noise might even tip the balance between two very different responses to the given clues, e.g., tip the balance between the "fight" or "flight" response to some shadowy form.

The effect of the independent superpositions of the "release" or "don't release" options, coupled with the uncertainty in the timing of the vesicle release at each of the trillions of nerve terminals will be to cause the quantum mechanical state of the brain to become a smeared out superposition of different macro-states representing different alternative possible plans of action. As long as the brain dynamics is controlled wholly by Process II — which is the quantum generalization of the Newtonian laws of motion of classical physics — all of the various alternative possible plans of action will exist in parallel, with no one plan of action singled out as the one that will actually occur. Some other process, beyond the local deterministic Process II, is required to pick out one particular real course of physical events from the smeared out mass of possibilities generated by all of the alternative possible combinations of vesicle releases at all of the trillions of nerve terminals. That other process is Process I, which brings in the action of the mind of the agent upon his brain.

This explanation of why quantum theory is pertinent to brain dynamics has focused on individual calcium ions in nerve terminals. That argument pertains to the *Process II component* of brain dynamics.

The equally important *Process I component* of the brain dynamics, which brings the mind of the agent into the dynamics, must be analyzed in terms of a completely different set of variables, namely certain *quasi-stable macroscopic degrees of freedom*. These specify the brain structures that enjoy the stability or persistence, and the causal connections needed to represent intentional actions and expected feedbacks.

The states of the brain that will be singled out by the actions P that specify the form of a Process I action will be more like the lowest-energy state of the simple harmonic oscillator discussed above, which tends to endure for a long time, or like the states obtained from such lowest-energy states by spatial displacements and shifts in velocity. Such states tend to endure as oscillating states, rather than immediately exploding. In other words, in order to get the needed stability properties the projection operators P corresponding to intentional actions should be constructed out of *oscillating states of macroscopic subsystems of the brain*, rather than out of sharply defined spatial states of the individual particles. The pertinent states will be functionally important brain analogs of a collection of oscillating modes of a drumhead, in which large collections of particles of the brain are moving in a coordinated way that will lead on to further coordinated activity.

In summary, the *need to use quantum theory* in brain dynamics arises from the dispersive quality of *Process II* action at the level of the ionic, and electronic, and atomic components of the brain. Hence *that* analysis is carried out at the individual-particle level. However, the opposing integrative and selective action, Process I, which brings in the mental (i.e., psychologically described) aspect, involves a completely different set of variables. Process I is specified by an operator P that singles out a quasi-stable large-scale pattern of brain activity that is the brain correlate of a particular mental intention.

It should be mentioned here that the actions P are *nonlocal*: they must act over extended regions, which can, and are expected to, cover large regions of the brain. Each conscious act is associated with a Process I action that coordinates and integrates activities in diverse parts of the brain. A conscious thought, as represented by the von Neumann Process I, effectively grasps as a whole an entire quasi-stable macroscopic brain activity.

11.7.2 Choices of the Process I Actions

It has been emphasized that the choices of which Process I actions actually occur are "free choices", in the sense that they are not specified by the currently known orthodox laws of physics. On the other hand, a *person's* intentions surely depend upon his brain. This means that we need to understand the process that determines the choice of P, which, within the framework of contemporary physical theory, is a free choice. In other words, the laws of contemporary quantum theory, although highly restrictive, are not the whole story: there is still work to be done. Hypotheses must be formulated and tested.

According to the theory, each experience is associated with the occurrence of a Process I event. As a simple first guess, let us assume, following a suggestion of Benjamin Libet and other psychologists, that the occurrence of a Process I action is triggered by a "consent" on the part of the agent, and that the rapidity with which consent is given can be increased by "mental effort".

To get a definite model, let $\{P\}$ be the set of actions P that correspond to possible mental intentions. Then let $P(t)$ be the "most probable P in $\{P\}$", where the probability is defined by brain state $S(t)$. In equations this most probable P in $\{P\}$ would be the P in $\{P\}$ that maximizes tr $PS(t)P/\text{tr } S(t)$.

The first hypothesis will be that the Process I event specified by $P(t)$ will occur if and only if a "consent" is given at time t.

To make mind efficacious it is assumed that "consent" depends on the *mental realities* associated with $P(t)$, and that "consent" can be given with a rapidity that is increased if the mental evaluation includes a feeling of effort. This simplest model makes the choice of the Process I action dependent both upon the physical state of the agent's brain, and also upon the mental realities associated with that action. It is assumed, here, that the consent associated with "hearing a nearby clap of thunder" is essentially passive: it will occur unless attention is strongly focused elsewhere. The important input of the mental aspect arises from the *effortful* focusing of mental attention that increases the rate at which consents are given.

Quantum theory *explains* how such a mental effort can strongly influence the course of brain events. Within the von Neumann framework this potentially very strong effect of mind upon brain is an automatic consequence of a well-known and well-studied feature of quantum theory called the quantum Zeno effect.

11.7.3 The Quantum Zeno Effect

If one considers only passive consents, then it is very difficult to identify any clean empirical effect of this intervention, apart from the production of low-level awareness. In the first place, the empirical averaging over the "Yes" and "No" possibilities tends to wash out all measurable effects. Moreover, the passivity of the mental process means that we have no independent self-controlled mental variable.

But the study of effortful and intentionally controlled attention brings in two empirically accessible variables, the intention and the amount of effort. It also brings in the important physical quantum Zeno effect. This effect is named for the Greek philosopher Zeno of Elea, and was brought into prominence in 1977 by the physicists Sudarshan and Misra.[11] It gives a name to the fact that repeated and closely spaced intentional acts can effectively hold the "Yes" feedback in place for an extended time interval that depends upon the *rapidity at which the Process I actions are happening*. According to our quantum model, this rapidity is controlled by the amount of effort being applied.

This quantum Zeno effect is, from a theoretical point of view, a very clean consequence of the von Neumann theory. It follows from the formula for the transition from the state PSP at time $t = 0$ to the state $(1 - P)S(t)(1 - P)$ at time t:

$$(1 - P)\,e^{-iHt}\,PSP\,e^{iHt}(1 - P) = \text{order } t \text{ squared.}$$

For small t the expression e^{iHt} becomes $1 + iHt +$ order t squared. Consequently, the terms of zeroth and first order in t on the left side of the above equation are both zero owing to the condition $P = PP$ on the projection operator P.

This result entails that by increasing sufficiently the rapidity of the Process I actions associated with a constant (or even slowly changing) operator P, an agent can keep the state S of his or her brain in the "Yes" subspace associated with states of the form $PS(t)P$.

This "holding-in-place" effect of rapidly repeated observations is known as the quantum Zeno effect, and is a macroscopic quantum effect in the conscious brain that is not diminished by the very strong interaction of the brain with its environment.

This result means that if a sequence of similar Process I events occur rapidly (on the time scale of the macroscopic oscillations associated with the associated actions P), then the "Yes" outcome can be held in place in the face of strong Process II mechanical forces that would tend to quickly produce the "No" feedback. Consequently, agents whose efforts can influence the rapidity of Process I actions would enjoy a survival advantage over competitors that lack this feature, for they could maintain beneficial activities longer than their Process I deprived competitors. This gives the leverage needed to link mind to natural selection, and also the leverage needed to allow us to link our mental intentions to our physical actions. For these efforts will then have intention-related physical effects, and his linkage can in principle be discovered, and integrated into behavior by the trial-and-error learning process mentioned earlier.

11.7.4 Support from Psychology

A person's experiential life is a stream of conscious experiences. The person's experienced "*self*" is part of this stream of consciousness: it is not an extra thing that is outside or apart from the stream. In James's words,

> *thought is itself the thinker*, and psychology need not look beyond.

The "self" is a slowly changing "fringe" part of the stream of consciousness. It provides a background cause for the central focus of attention.

The physical brain, evolving mechanically in accordance with the local deterministic Process II does most of the necessary work, without the intervention of Process I. It does its job of creating, on the basis of its interpretation of the clues provided by the senses, a suitable response. But, owing to its quantum nature, the brain necessarily generates an amorphous mass of overlapping and conflicting templates for action. Process I acts to extract from this jumbled mass of possibilities a dynamically stable configuration in which all of the quasi-independent modular components of the brain act together in a maximal mutually supportive configuration of nondiscordant harmony that tends to prolong itself into the future and produce a characteristic subsequent feedback. This is the preferred "Yes" state PSP that specifies the form of the Process I event. But the quantum rules do not assert that this preferred part of the prior state S necessarily comes into being: they assert, instead, that if this process is activated—say by some sort of "consent"—then this "Yes" component PSP will come into being with probability tr PSP/tr S.

The rate at which consents are given is assumed to be increasable by mental effort.

The phenomenon of "will" is understood in terms of this effortful control of Process I, which can, by means of the quantum Zeno effect, override strong mechanical forces arising from Process II, and cause a large deviation of brain activity from what it would be if no mental effort were made.

Does this quantum-physics-based conception of the connection between mind and brain explain anything in the realm of psychology?

Consider some passages from *Psychology: The Briefer Course*, written by William James. In the final section of the chapter on attention James writes:

> I have spoken as if our attention were wholly determined by neural conditions. I believe that the array of things we can attend to is so determined. No object can catch our attention except by the neural machinery. But the amount of the attention which an object receives after it has caught our attention is another question. It often takes effort to keep mind upon it. We feel that we can make more or less of the effort as we choose. If this feeling be not deceptive, if our effort be a spiritual force, and an indeterminate one, then of course it contributes coequally with the cerebral conditions to the result. Though it introduce no new idea, it will deepen and prolong the stay in consciousness of innumerable ideas which else would fade more quickly away.[12]

In the chapter on will, in the section entitled "Volitional effort is effort of attention",[13] James writes:

> Thus we find that *we reach the heart of our inquiry into volition when we ask by what process is it that the thought of any given action comes to prevail stably in the mind.*

and later:

> The essential achievement of the will, in short, when it is most "voluntary", is to attend to a difficult object and hold it fast before the mind ... Effort of attention is thus the essential phenomenon of will.

Still later, James says:

> *Consent to the idea's undivided presence, this is effort's sole achievement* . . .
> Everywhere, then, the function of effort is the same: to keep affirming and
> adopting the thought which, if left to itself, would slip away.

This description of the effect of mind on the course of mind–brain process is remarkably in line with *what had been proposed independently from
purely theoretical considerations of the quantum physics of this process*. The
connections specified by James are *explained* on the basis of the same dynamical principles that had been introduced by physicists to explain atomic
phenomena.

In the quantum theory of mind-brain being described here there are
two separate processes. First, there is the unconscious mechanical brain
process called Process II. As discussed at length elsewhere in the present
book (page 124), this brain processing involves dynamical units that are
represented by complex patterns of neural activity (or, more generally, of
brain activity) that are "facilitated" (i.e., strengthened) by use, and are such
that each unit tends to be activated as a whole by the activation of several
of its parts. The activation of various of these complex patterns by cross-
referencing—i.e., by activation of several of its parts—coupled to feedback
loops that strengthen or weaken the activities of appropriate processing
centers appears to account for the essential features of the mechanical part
of the dynamics in a way that in many cases is not greatly different from that
of a classical model, except for the creation of a superposition of a host of
parallel possibilities that according to the classical concepts could not exist
simultaneously.

The second process, von Neumann's Process I, is a selection process
that is tied to intentions, and that is needed in order to separate what is
experienced from the continuum of alternative possibilities generated by
Process II.

An extended discussion of nontrivial agreement of these features with
a large body of recent data from the field of the psychology of attention is
described elsewhere.[14]

11.8 Quantum Theory in Neuroscience

Scientists in different fields are to some extent free to choose what sort of
models or theories they use to organize, explain, understand, and predict the
observed features of the data in their field, and to guide their further inquiries.
On the other hand, the ideal of the unity of science gives precedence to

models that mesh with the basic principles of physics, or at least do not contradict them.

On the basis of that ideal the quantum theoretical framework would seem to be superior to the classical one for explaining correlations between psychologically and physically described data. It not only accommodates — and arises from — an adequate account of the physical and chemical processes that underlie brain behavior, but also provides a theoretical framework that has places for the two kinds of data that need to be brought into theoretical concordance, and it also specifies theoretical conditions on the two-way causal connection between these two kinds of data. The concepts of classical physics, on the other hand, are not only known to be inadequate to deal with, for example, the dynamics of ionic motions, but have no natural place for psychologically described data, and no capacity to explain the apparent causal efficacy of willful effort, except as a mysterious illusion arising in connection with conscious realities that are conceptually alien to the concepts of classical physics. Moreover, the causal efficacy of willful effort is eliminated by the approximation that produces classical physics.

To bring these theoretical ideas down to the practical level let us consider the experiments of Ochsner et al.,[15] with particular attention to the following four key questions (posed by neuroscientist Mario Beauregard):

1   How does the quantum mechanism work in this case, in comparison to what the classical account would say?
2   How do we account for the rapid changes occurring in large neural circuits involving millions of neurons during conscious and voluntary regulation of brain activity?
3   How does consciousness "know" where and how to interact in the brain in order to produce a specific psychological effect?
4   Is consciousness localized, and, if so, how and in what sense; or does it lie, instead, "outside of space"?

Reduced to their essence, the experiments in question consist first of a training phase in which the subject is taught how to distinguish, and respond differently to, two instructions given while viewing emotionally disturbing visual images: ATTEND (meaning passively "be aware of, but do not try to alter, any feelings elicited by") or REAPPRAISE (meaning actively "reinterpret the content so that it no longer elicits a negative response"). The subjects then perform these mental actions during brain data acquisition. The visual stimuli, when passively attended to, activate limbic brain areas and when actively reappraised activate prefrontal cerebral regions. [The succinct formulation given in this paragraph is due mainly to Dr Jeffrey Schwartz.]

From the classical materialist point of view this is essentially a conditioning experiment, where, however, the "conditioning" is achieved via linguistic access to cognitive faculties. But how do the cognitive realities involving "knowing," "understanding", and "feeling" arise out of motions of the miniature planetlike objects of classical physics, which have no trace of any experiential quality? And how do the vibrations in the air that carry the instructions get converted into feelings of understanding? And how do these feelings of understanding get converted to effortful actions, the presence or absence of which determine whether the limbic or frontal regions of the brain will be activated.

Within the framework of classical physics these connections between feelings and brain activities are huge mysteries. The classical materialist claim is that someday these connections will be understood. But the basic question is whether these connections will ever be understood in terms of a physical theory that is known to be false, and that, moreover, results from an approximation that, according to contemporary physical theory, systematically excludes the effect of psychological realities upon physiological realities that these neuropsychology experiments reveal. Or, on the other hand, will the eventual understanding of this linkage accord with causal linkage between mental realities and brain activities that orthodox (von Neumann) contemporary physical theory entails.

There are important similarities and also important differences between the classical and quantum explanations of the experiments of Ochsner et al. In both approaches the particles in the brain can be conceived to be collected into nerves and other biological structures, and into fluxes of ions and electrons, which can all be described reasonably well in essentially classical terms. However, in the classical description the dynamics is well described in terms of the local deterministic classical laws that govern these classical quantities, insofar as they are precisely defined.

Quantum theory asserts, however, that the condition that these classical quantities be *precisely defined* is unrealistic: Heisenberg's uncertainty principle asserts that this assumption is not justified: one must accept at least some small amount of cloudlike uncertainty. But small uncertainties rapidly grow into larger uncertainties. The discussion of the ionic motions in nerve terminals exemplifies this growth of uncertainty: the state of the brain rapidly fans out into a state that encompasses many possible experiential states.

This incursion into the dynamics of growing uncertainties renders the classical approach basically incomplete: it can never lead to well-defined experiential states, except by actually violating the quantum uncertainty principle.

There is a well-known and powerful process in quantum theory that strongly influences this expansion of the state of the brain into a state that encompasses many alternative experiential possibilities. It is called "environmental decoherence". The interactions of the brain with its environment rapidly reduce the state S of the brain into what is called a "mixture". This means that the interference effects between significantly different classically describable possibilities become markedly attenuated. That effect is, however, already completely accounted for in the von Neumann state S of the brain: environmental decoherence is describable within von Neumann's formulation, and it in no way upsets or modifies the von Neumann theory described here. Indeed, it makes quantum theory more accessible to neuroscientists by converting the complex mathematical concept of a quantum state into a structure that can be visualized as simply a smear of virtual classically conceived states: the quantum state of the brain is effectively transformed by environmental decoherence effects into a continuous smear of classically describable potentialities that becomes converted to a rapid sequence of discrete experiential realities by Process I actions. Thus the quantum brain dynamics becomes much easier to conceive and to describe because the environmental decoherence effect allows classical language and imagery to be validly used in an important way. But environmental decoherence has never been shown to obviate the need for von Neumann's Process I.[16]

One could, despite violating the quantum laws, try to pursue a quasi-classical calculation. This would be a classical-type computation with the quantum-mandated uncertainties folded in as probability distributions, and with certain classically describable brain states identified as the "neural correlates" of the various possible experiential states. One could then produce, in principle, the same general kinds of statistical predictions that quantum theory would give.

This sort of quasi-classical approach would, in fact, probably give results very similar to quantum theory for situations arising from "passive attention". For in these cases mind is acting essentially as a passive witness, in a way that is basically in line with the ideas of classical physics.

But quantum theory was designed to deal with the other case, in which the conscious action of an agent—to perform some particular probing action—enters into the dynamics in an essential way. Within the context of the experiments by Ochsner et al., quantum theory provides, via the Process I mechanism, an explicit means whereby the mental effort actually causes—by catching and actively holding in place—the prefrontal activation instead of the limbic one. Thus, within the quantum framework, the causal relationship between the mental effort and the observed brain changes is *dynamically accounted for*. Analogous quantum mechanical reasoning can be utilized to explain the data of Beauregard[17] and related studies of self-directed neuroplasticity[18].

The second question is: How do we account for the rapid changes induced by mental effort in large brain circuits?

The answer is that the nonlocal operator P that represents the intention singles out a large quasi-stable and functionally important brain state that is likely to produce the expected feedbacks. Large functionally effective brain activities are singled out and linked to mental effort through learning, which depends upon the fact that the mental efforts, per se, have physical consequences. These discrete macroscopic functional states are singled out from the smear of possibilities by the nonlocal Process I. Thus quantum theory describes the mathematical machinery that links the mentalistically described intention to the physically described macroscopic state of the brain that implements it.

The third question is: How does consciousness "know" where and how to interact in the brain in order to produce a specific psychological effect?

The answer is that felt intentions, per se, have physical consequences, and thence experiential consequences. Hence an agent can learn, by trial and error, how to select an intentional action that is likely to produce a feedback that fulfills that intention.

The fourth question is: Is consciousness localized, and, if so, how and in what sense; or does it lie, instead, "outside of space"?

Each conscious event is associated with a Process I action that involves an action P that is necessarily nonlocal, for mathematical reasons. Moreover, the "Yes" part must have the functional properties needed to set in motion the brain–body activity that is likely to produce the intended feedback experience. Thus each conscious action would, in order to meet these requirements, act over some functionally characterized extended portion of the brain. (In fact, for reasons that go well beyond the scope of this article, this event also induces effects in faraway places: these effects are the causes, within the von Neumann ontology, of the long-range nonlocal effects associated with the famous theorem of John Bell.[19])

11.9 Ramifications in Neuroscience

The situations in neuroscience and atomic science are similar. Owing to the Heisenberg uncertainty principle, micro-properties such as the velocities of the ions emerging from narrow ion channels are in principle unknowable. Thus the computation of the causal behavior of a conscious brain is in principle impossible. Thus just as in atomic physics, and indeed as a direct consequence of the basic principle of atomic physics, there is both room for, and, at least at the practical level, a rational need for, the input of psychologically described data that can according to quantum theory be rationally treated as replacements for the accessible-in-principle micro-properties. According to orthodox quantum theory, the micro-properties postulated by classical physical theory do not exist, but the dynamical gap created by their absence can be partially filled by accepting the psychologically describable and partially controllable data pertaining to conscious human choices about how to act as primary data describing pragmatically independent realities.

The breakdown *in principle* of the possibility of a complete bottom–up micro-local causal description opens the door to the quantum psychophysical description, which consistently combines the bottom–up micro-locally determined Process II with the top–down mentally controlled Process I.

Francis Crick and Christoff Koch have published recently in *Nature Neuroscience* a commentary entitled "A framework for consciousness".[20] They explain that their framework will "not have rigid laws as physics does". But they put forth a ten-fold "point of view for an attack on" the scientific problem of consciousness. Much of their proposal focuses on neuro-anatomical details. But the general features of their framework are in very good agreement with the quantum psychophysical framework described in the present volume.

Crick and Koch explain that they are, in this initial phase of their program, restricting themselves to "attempting to find the neural correlates of consciousness (NCC), in the hope that when we can explain the NCC in causal terms, this will make the problem of qualia clearer". But what does a causal account dealing only with the neural correlates of consciousness say about the causal properties of the conscious realities themselves?

1   The (unconscious?) homunculus. Crick and Koch speak of the "overwhelming illusion" of the existence of a conscious homunculus, and suggest that this illusion may "reflect in some way the general organization of the brain". But how do they conclude that the overwhelming intuition that our thoughts can influence our actions is an illusion? The only basis for that allegation is the known-to-be-false classical physical

theory. What is the rational basis for denying the validity of this over-whelming intuition, rather than denying the validity of that provably false theory, and accepting, instead, the validated physical theory that validates this overwhelming intuition?

2   Zombie modes and consciousness. Crick and Koch say

> Consciousness deals more slowly with . . . and takes time to decide on ap-propriate thoughts and actions.

But how can consciousness, or conscious decisions, deal with anything if only their neural correlates are considered. Some property beyond mere correlation is needed for consciousness to be able to deal with anything, or to decide on actions. The quantum psychophysical theory justifies this causal language.

3   Coalitions of neurons. Crick and Koch say that the winning coalition "embodies what we are conscious of" and "produces consciousness". But how does a coalition "produce" consciousness, within the frame-work of classical physics? All that can ever be derived or deduced from the principles of classical physics are combinations of simple math-ematical properties imbedded in spacetime, and functional properties deducible from them. The concept of "producing consciousness" is not part of classical physics. If one wants to argue that this "production of consciousness" property is an ontological aspect of the classically conceived world that simply is not specified or captured by the classical principles, then there is the difficulty that there can be no ontologi-cal reality that is even *compatible* with the classical principles. Is it not, therefore, more rational to accept the theory that quantum physi-cists have already discovered, and extensively studied and verified, and which, in its orthodox formulation, brings consciousness into the the-ory in a rationally coherent, causally efficacious, and practically useful manner?

4   Snapshots. Crick and Koch say

> We propose that conscious awareness (for vision) is a series of static snap-shots, with "motion" painted onto them.
> Perception occurs in discrete epochs.

This refers to "awareness" and "perception", but presumably it must be the NCC that have these discrete epochs. But dynamical discreteness is incompatible with classical physics. However, a series of discrete conscious events is exactly what quantum theory gives.

5   Attention and binding. Crick and Koch say:

> Attention can usefully be divided into two forms: either rapid, saliency driven, and bottom–up or slower, volitionally controlled, and top–down.

The quantum approach *explains* the occurrence of these two kinds of attention, and also binding, as a consequence of the basic laws of physics. The micro-causal Process II is high-speed, saliency-driven, and parallel, whereas the nonlocal, integrative, and effortfully deliberative Process I consists of a *series* of similar actions held in place by the quantum Zeno effect.

The quantum psychophysical theory of the conscious brain is, like quantum theory in general, a *pragmatic theory*. It is set within the framework of communicable descriptions of our intentional actions, and the experiential feedbacks that result from these actions. It justifies *dynamically* our intuition that our psychologically described mental efforts are able to influence our mental and physical behavior in the way that we feel they do. Thus science becomes intelligible: our physical communications are allowed to convey the real knowledge, information, instructions, and meanings that they do in fact carry. They do the job of communicating physically efficacious ideas, rather than being physical vibrations that encode instructions passing between complex biological computers that mysteriously produce, in some presently (and surely eternally) incomprehensible mechanical way, the *illusion* that our thoughts are doing what we think they are doing.

But why should neuroscience bind itself to this essentially seventeenth-century approach based on logically inadequate principles and known-to-be-nonexistent entities when contemporary physical theory provides a rationally coherent alternative that accords with all the new and old physics data, and brings consciousness into the theory at the foundational level, in tight mathematically controlled coordination with the physically described brain.

Shifting to the quantum psychophysical approach to the mind–brain problem means switching to a new research posture. The objective is no longer to explain how a classically conceived brain can "produce" or "be" psychologically experienced consciousness. It is rather to elucidate the respective roles of the physically described brain and psychologically described mind in the determination of the content and timings of the stream of conscious Process I actions.

In summary: Neuropsychological theory is greatly simplified by accepting the fact that brains must in principle be treated quantum mechanically. Accepting that obvious fact means that the huge deferred-to-the-future question of how mind is connected to a classically described brain must, in principle, be replaced by the already partially resolved question of how mind is connected to a quantum mechanically described brain. That shift means adopting the *same* pragmatic solution that atomic physicists adopted when faced with this same problem of accounting coherently for the effects of *mentalistically described human intentional actions upon the physically described systems that those actions act upon.* The benefits of adopting the pragmatic quantum approach may be as important to progress in neuroscience as they were in atomic physics.

References

1 I. Newton, *Principia Mathematica* (1687) [*Newton's Principia*, edited by Florian Cajori (University of California Press, Berkeley, 1964), p. 634].

2 W. James, *The Principles of Psychology* (Dover, New York, 1950; reprint of 1890 text), vol. 1, p. 138 (italics are mine).

3 Ref. 2, p. 136 (italics are mine).

4 R. W. Sperry, Turnabout on Consciousness: A Mentalist View, *J. Mind & Behavior* **13**, 259–280 (1992), p. 276.

5 N. Bohr, *Atomic Physics and Human Knowledge* (Wiley, New York, 1958), p. 81; *Essays 1958/1962 on Atomic Physics and Human Knowledge* (Wiley, New York, 1963), p. 15.

6 J. von Neumann, *Mathematical Foundations of Quantum Theory* (Princeton University Press, Princeton, 1955), p. 418.

7 N. Bohr, *Atomic Physics and Human Knowledge* (Wiley, New York, 1958), p. 73.

8 Ref. 6, p. 421.

9 W. Heisenberg, The Representation of Nature in Contemporary Physics, *Daedalus* **87**, 95–108 (1958).

10 M. Cataldi, E. Perez-Reyes, and R. W. Tsien, Difference in Apparent Pore Sizes of Low and High Voltage-Activated Ca^{2+} Channels, *J. Biol. Chem.* **277**, 45969–45976 (2002).

11 B. Misra and E. C. G. Sudarshan, The Zeno's Paradox in Quantum Theory, *J. Math. Phys.* **18**, 756–763 (1977).

12 W. James, *Psychology: The Briefer Course*, in *William James: Writings 1879–1899* (Library of America, New York, 1992), p. 227.

13 Ref. 12, p. 417.

14 H. P. Stapp, Quantum Theory and the Role of Mind in Nature, *Found. Phys.* **11**, 1465–1499 (2001).

15 K. N. Ochsner, S. A. Bunge, J. J. Gross, and J. D. E Gabrieli, Rethinking Feelings: An fMRI Study of the Cognitive Regulation of Emotion, *J. Cog. Neurosci.* **14:8**, 1215–1229 (2002).

16 H. P. Stapp, The Basis Problem in Many-Worlds Theories, *Can. J. Phys.* **80**, 1043–1052 (2002). H. P. Stapp, A Bell-Type Theorem without Hidden Variables (2003), to appear in *Amer. J. Phys.*

17 M. Beauregard, J. Levesque, and P. Bourgouin, Neural Correlates of the Conscious Self-Regulation of Emotion', *J. Neurosci.* **21**, RC165, 1–6 (2001).

18 J. M. Schwartz and S. Begley, *The Mind and the Brain: Neuroplasticity and the Power of Mental Force* (HarperCollins, New York, 2002).

19 J. Bell, On the Einstein–Podolsky–Rosen Paradox, *Physics* **1**, 195–201 (1964). See ref. 16.

20 F. Crick and C. Koch, *Nature Neurosci.* **6(2)**, 119–126 (Feb. 2003).

12 Societal Ramifications of the New Scientific Conception of Human Beings

A major revolution occurred in science during the twentieth century. This change leads to a profound transformation of the scientific conception of human beings. Whereas the former conception of man undermines rational moral philosophy, the new one can buttress it.

I intend to explain here this tectonic shift in science, and its relevance to our lives.

I begin by listing three huge turnabouts in science that occurred during the past four centuries. I shall describe how each of them radically transformed our scientific understanding of human beings, and will then spotlight the moral, social, and philosophical significance of these developments. I then conclude by describing practical measures for promoting a rapprochement of science and moral philosophy.

The first of the three great shifts was the creation of what is called "classical physics". This development was initiated during the seventeenth century by Galileo, Descartes, and Newton, and was completed early in the twentieth century by the inclusion of Einstein's theories of special and general relativity.

The second major shift was the creation of quantum theory. This revision began at the outset of the twentieth century with Max Planck's discovery of the quantum of action, and was completed in the years 1925 to 1927, principally by Heisenberg, Bohr, Pauli, Dirac, Schrödinger, and Max Born.

The third crucial shift was the integration of the mental and physical aspects of nature. It was begun in the early 1930s by John von Neumann and Eugene Wigner, and has developed rapidly during the past decade.

Each of these three developments has a main theme.

The main theme of classical physics is that we live in a clocklike universe, and that even our bodies and brains are mechanical systems. The theory asserts that nature has a "material" part that consists of tiny localized bits of matter, and that every motion of each of these minute material elements is completely determined by contact interactions between adjacent material elements. This material part of nature includes our bodies and our brains. Hence, according to classical physics, each of our bodily actions is completely fixed by mechanical processes occurring at atomic or subatomic levels.

Classical physics accommodates the existence also of another part of nature, which consists of our human thoughts, ideas, feelings, and sensations. However, the existence of these experiential aspects of nature is not entailed by the principles that govern the behavior of material parts. The classical-physics framework, which purports to specify completely the motion of every bit of matter, contains no requirement for any experiential aspect of nature to exist at all: the principles of classical physics fail to entail the existence of the defining characteristics of experiences, namely the way that they feel. Since, within the classical framework, our experiences need not even exist, they cannot, within that framework, be the causes of any physical action: our thoughts are reduced to at most passive bystanders. They are not elements of the chain of events that are, within that theory, the necessary and sufficient causes of every material motion, and hence of every bodily action.

This causal irrelevance of our thoughts within classical physics constitutes a serious deficiency of that theory, construed as a description of reality. Such an inertness of thoughts, if it were actually true, would mean that reality has experiential parts that have no logically required dynamical link to the physical world that the theory describes: nature would be split into two effectively independent parts.

Such a separation is philosophically repugnant. But, besides that, it fails to explain your direct knowledge that you can, by your willful effort, cause your thumb to move. No such effect of mind on matter is explained by the supposedly causally complete classical physics: the felt effectiveness of your thoughts in influencing your bodily actions becomes merely a strange illusion. But how can a rationally coherent moral philosophy be based on a conception of nature in which the thoughts of a normal human being have no effect upon what he does?

This difficulty has, quite rightly, been the topic of intense philosophical interest and effort for over three hundred years, but no satisfactory explanation has been found.

A second problem with this classical-physics conception of man is the difficulty in understanding the close correlation between brain process and conscious process in the context of the evolution of our species: if there were no causal feedback from conscious process to brain process, then creatures with normal mind–brain correlations would be no better off than organisms with totally disconnected minds and brains. Natural selection would not favor creatures whose ideas about where food is located are correlated to where food is actually located over creatures that always think food is behind them.

During the twentieth century this classical theory of nature was found to be incompatible with the emerging empirical data pertaining to the detailed properties matter. A new approach, called quantum theory, was devised. It explains both all the empirical facts that are explained by classical physics, plus all of the newer experimental data in which the classical predictions fail.

The new theory differs profoundly from its predecessor. Classical physics was a deterministic theory about postulated localized bits of matter, whereas quantum theory is a probabilistic theory about nonlocalized bits of information.

This great step forward was initially bought at a heavy price: scientists had to renounce, in principle, their traditional goal of seeking the "truth" about what was going on in the physical world. They were forced to retreat to the position of being satisfied with a set of practical rules that allowed them to make statistical predictions about connections between their empirical observations, renouncing all claims to any understanding of what was actually going on.

This essentially subjective approach to physical theory was devised and promulgated by Niels Bohr, and the physicists that he gathered about him in Copenhagen. Hence it is known as the "Copenhagen interpretation". It works exceedingly well in actual practice.

In spite of the unparalleled practical success of the restricted program, some scientists have been unwilling to abandon the ideal that science should strive to find a rationally coherent conception of the reality that lies behind the empirical facts.

The only successful effort in this direction that I know of is the one initiated by John von Neumann and Eugene Wigner. It accepts as real the subjective elements of experience that are the basic elements of Copenhagen quantum theory, and relates them to an equally real, but nonmaterial, objective physical universe.

Under the impetus of the rapidly growing scientific interest in the connection between the objective and subjective aspects of nature the von Neumann–Wigner approach has been developed over the past decade into a post-Copenhagen quantum theory that explains a great deal of the detailed structure of the emerging data in this field. This development allows quantum theory to be elevated from a set of practically successful — but mysterious — rules, to a rationally coherent conception of man and nature.

The basic theme of both Copenhagen and post-Copenhagen quantum theory is that the physical world must be understood in terms of information:

the "tiny bits of matter" that classical physics had assumed the world to be built out of are replaced by spread-out nonmaterial structures that combine to form a new kind of physical reality. It consists of an objective carrier of a growing collection of "nonlocalized bits of information" that are dynamically related to experiential-type realities.

Each subjective experience injects one bit of information into this objective store of information, which then specifies, via known mathematical laws, the relative probabilities for various possible future subjective experiences to occur. The physical world thus becomes an evolving structure of information, and of propensities for experiences to occur, rather than a mechanically evolving mindless material structure. The new conception essentially fulfills the age-old philosophical idea that nature should be made out of a kind of stuff that combines in an integrated and natural way certain mindlike and matterlike qualities, without being reduced either to classically conceived mind or classically conceived matter. This new quantum structure entails the validity of all the scientifically validated empirical data, while at the same time explaining how our thoughts can influence our actions in a way concordant with our normal experience of that connection.

Another pertinent property of the new theory concerns "locality".

Classical dynamics is "local" in the sense that all causation is via contact interaction between neighboring bits of matter. Von Neumann–Wigner quantum theory violates that condition in two different ways. The first pertains to the mechanism by which a person's thoughts influence his actions. That process is not a local process in which tiny elements act upon their neighbors. It is a process involving bits of information that reside in space-time structures that can extend over large portions of the person's brain or body, and that are associated with whole experiences. Von Neumann has given a name to this important nonlocal process: he calls it Process I.

There is also a second way in which the action of subjective experiences upon the physical world turns out to be "nonlocal": what a person decides to do in one place can instantly influence what is true in distant places. That feature seems, on the face of it, to contradict the theory of relativity, which forbids sending signals faster than light. However, quantum theory is exquisitely constructed so that all of the empirically testable consequences of the theory of relativity are preserved. But, in spite of this restriction, the picture of nature that emerges is one in which the global evolution of the universe is controlled in part by choices made by localized agents, such as human beings. The causal roots, or origins, of these choices are not specified by any laws that we yet know or understand. In that very specific sense these choices are "free". However, they can affect the behavior of the agent himself, and necessarily have, moreover, effects on faraway physical events.

What are the moral, social, and philosophical implications of this profound revision of our scientific understanding of man and nature?

There has been a long-standing conflict between classical physics and rational moral philosophy: according to the precepts of classical physics each man is a machine ruled by local material processes alone, whereas rational moral philosophy is based on the presumption that what a normal human being knows and understands can make a difference in how he behaves. Jurisprudence is, accordingly, based on the premise that insofar as a person was able to know the nature and quality of the act he was doing or to know he was doing what was wrong, then he is responsible for that act.

This rule is based on the premise that knowing and understanding can influence behavior. But classical physics, by claiming all behavior to be completely determined by atomic or subatomic processes that do not entail the existence of knowing or understanding, undermines that premise. It would make no sense to make responsibility hinge on knowing and understanding if knowing and understanding cannot influence action. One must place responsibility where power lies.

Quantum theory, unlike classical physics, allows a person's mental process to make a difference in how his body behaves. Von Neumann quantum theory injects human thoughts into the causal structure of nature in an irreducible way that allows a person's mental effort to influence his bodily actions. This influence of mind is not just a redundant re-expression of other known or postulated laws, but is an effect that has no other known cause.

The situation is this: Quantum theory dynamics is like "twenty questions". First some definite question with a Yes or No answer is chosen. Then nature delivers an answer, Yes or No. The relative probabilities of the two possible answers, Yes and No, are specified by the theory, and are therefore not controllable by human beings.

But both Copenhagen and post-Copenhagen quantum theory allow an "agent" to choose which question will be asked. These choices are, in general, not specified by any known laws of physics. They are in this very specific sense "free choices".

But these choices can, according to the known laws of quantum theory, influence the physical behavior of the agent. Thus twenty-first-century science, unlike nineteenth-century science, does not reduce human beings to mechanical automata, deluded by the scientifically unsupportable belief that their thoughts can make a difference in how they behave. Rather it elevates human beings to agents whose "free choices" can, according to the known laws, actually influence their behavior.

The problem with classical physics is not just some airy philosophical abstraction. The philosophical dilemma has trickled down into the workings of our society. The Australian supreme court justice David Hodgson has written a book, *The Mind Matters*,[1] that documents the pervasive and pernicious effect that the idea that "mind does not matter" is having upon our legal system.

An example occurred in San Francisco: Dan White walked into the office of Mayor George Moscone and shot him dead, and then walked down the hall and shot dead Supervisor Harvey Milk. White got off with five years, on the basis of the infamous "twinkie defense" that he was not responsible for his actions, owing to derangement caused by junk foods.

One of the most influential philosophers of the present time, Daniel Dennett, argues in his book *Consciousness Explained*,[2] and elsewhere, that our conscious thoughts, as we normally understand them, do not exist, and ought to be drummed out of our scientific understanding of human beings. He explained his basic motivation:

> . . . a brain was always going to do what it was caused to do by current local mechanical circumstances.[3]

If this claim were indeed true, then Dennett's conclusions might be valid. But the clear message of the quantum theory is that Dennett's assumption is not valid: what a person's brain does can, according to the quantum theory, be strongly influenced by a nonlocal causal process connected to the person's conscious choices and mental efforts. Consciousness can play a nonredundant causal role in the determination of our actions: it can play the very role that we intuitively feel that it plays. Quantum theory allows your mind and your brain to co-author your physical actions.

A central moral issue concerns "values".

What a person values depends, basically, on what he believes himself to be. If he believes that he is an isolated hunk of protoplasm, struggling to survive in a hostile world, or a physical organism constructed by genes to promote their own survival, then his values will tend to be very different from those of a person who regards himself as a being with a mindlike aspect that makes conscious choices that control in part his own future, and are also integral parts of the global process that generates the unfolding of the universe.

The second half of the twentieth century featured the rise of postmodernism. It denies the relationship between discourse and reality, and claims that "what we think we know" is just "what we have been discursively disciplined to believe". This abandonment of the idea of objective truth leads directly to moral relativism. It draws support both from the theory of relativity, which proclaims that what is true about nature depends upon the observer, and from the Copenhagen philosophy that renounces, even in science, the search for objective truth.

But post-Copenhagen quantum theory sees the Copenhagen rejection of all inquiry about the nature of reality as merely a transitory phase between the old classical conception of reality to a more unified contemporary conception of nature. But this profound shift in what science says about the nature of the physical world, and of human beings, has yet to sink into the public consciousness.

One thing that needs to be done to resuscitate moral philosophy is to infuse into the intellectual milieu an awareness of the important relevant changes wrought by quantum theory in our understanding of the nature of man.

This initiative would involve the introduction into curricula, at all levels, of the contemporary quantum conception of nature in terms of information. False mechanistic ideas inculcated into tender minds at an early age are hard to dislodge later. If our children are taught that the world is a machine built out of tiny material parts, then both science and philosophy are damaged. The progress of science is inhibited by imbuing young minds with an incorrect idea of the nature of reality, and the pernicious philosophical idea that man is made of classically conceived matter is not exposed as being incompatible with the empirical facts.

One might think that the ideas of quantum physics are too counterintuitive for young minds to grasp. Yet students have no trouble comprehending the even more counterintuitive classical idea that the solid chairs upon which they sit are mostly empty space. Children and students who, through their computers, deal all the time with the physical world conceived of as a repository and transmitter of information should grasp far more easily the quantum concept of the physical world as a storehouse and conveyor of information than the classical concept of physical reality as a horde of unseen particles that can somehow be human experience. A thoroughly rational concept into which one's everyday experiences fit neatly should be easier to comprehend than a seventeenth-century concoction that has no place for one's own being as an active agent with efficacious thoughts, a concoction that has consequently confounded philosophers from the day it was invented, and which has now pushed some philosophers to the extremity of trying to convince us that consciousness, as we intuitively understand it, does not exist, or is an illusion, and other philosophers to the point of making truth a purely social construct.

In order to free human beings from the false materialist mind-set that still infects the world of rational discourse, a serious effort is needed to move people's understanding of what science says out of the seventeenth century and into the twenty-first.

One problem stands in the way of pursuing this updating of the curricula. Most quantum physicists are interested more in applications of quantum theory than in its ontological implications. Hence they often endorse the "Copenhagen" philosophy of renouncing the quest to understand reality, and settling, instead, for practical rules that work. This forsaking by physicists of their traditional goal of trying to understand the physical world means that there is now no official statement as to the nature of reality, or of man's place within it. Still, I believe that there will be near-unanimous agreement among quantum physicists that, to the extent that a rationally coherent conception of physical reality is possible, this reality will be informational in character, not material. For the whole language of the quantum physicist, when he is dealing with the meaning of his symbols, is in terms of information, which an agent may or may not choose to acquire, and in terms of Yes-or-No answers that constitute bits of information. Just getting that one idea across could make a significant inroad into the corruptive materialist outlook that, more than three-quarters of a century after its official demise as a basic truth about nature, still infects so many minds.

References

1 D. Hodgson, *The Mind Matters: Consciousness and Choice in a Quantum World* (Oxford University Press, Oxford, 1991).
2 D. C. Dennett, *Consciousness Explained* (Little, Brown, and Company, Boston, 1991).
3 D. C. Dennett, in *A Companion to the Philosophy of Mind*, edited by Samuel Guttenplan (Blackwell, Oxford, 1994), p. 237.

13 Physicalism Versus Quantum Mechanics

13.1 Introduction

The widely held philosophical position called "physicalism" has been described and defended in a recent book by Jaegwon Kim.[1] The physicalist position claims that the world is basically purely physical. However, "physical" is interpreted in a way predicated, in effect, upon certain properties of classical physics that are contradicted by the precepts of orthodox quantum physics. Kim's arguments reveal two horns of a dilemma that the physicalist is forced to face as a consequence of accepting this classical notion of "physical". Kim admits that neither of the two options, "epiphenomenalism" or "reduction", is very palatable, but he finds a compromise that he deems acceptable.

The central aim of this chapter is to show that the physicalist's dilemma dissolves when one shifts from the classical notion of the physical to the quantum mechanical notion. Understanding this shift involves distinguishing the classical notion of the mind–brain connection from its quantum successor.

To make clear the essential features of the quantum mechanical conception of the mind–brain connection, I shall describe here a model that is a specific realization of a theory I have described in more general terms before.[2] Being specific reduces generality, but having a concrete model can be helpful in revealing the general lay of the land. Also, the specific features added here resolve in a natural way the puzzle of how our descriptions of our observations can be couched in the language of classical physics when our brains are operating, fundamentally, in accordance with the principles of quantum theory. The specific model also shows how the thoroughly quantum mechanical (quantum Zeno) effect, which underlies the power of a person's conscious thoughts to influence in useful ways the physically described processes occurring in that person's brain, is not appreciably disrupted either by "environmental decoherence" effects or by thermal effects arising from the "hotness" of the brain.

In order to communicate with the broad spectrum of scientists and philosophers interested in the connection between mind and brain, I will review in the following section the historical and conceptual background of the needed quantum mechanical ideas, and then describe an approach to the mind–body problem that is based fundamentally on quantum theory, but that adds several specific new ideas about the form of the mind–brain connection.

13.2 Quantum Mechanics and Physicalism

Rather than just plunging ahead and using the concepts and equations of quantum mechanics, and thereby making this work unintelligible to many people that I want to reach, I am going to provide first an historical and conceptual review of the extremely profound changes in the philosophical and technical foundations that were wrought by the transition from classical physics to quantum physics. One key technical change was the shift from the numbers used in classical mechanics to describe properties of physical systems to the associated *operators* or *matrices* used to describe related actions. This technical shift emerged, unsought, from a seismic conceptual shift. Following the path blazed by Einstein's success in creating special relativity, Heisenberg changed course. Faced with a quarter century of failures to construct a successful atomic theory based upon the notion of some presumed-to-exist spacetime structure of the atom, Heisenberg attempted to build a theory based upon our observations and measurements, rather than upon conjectured microscopic spacetime structures that could be postulated to exist, but that were never directly observed or measured. This shift in orientation led to grave issues concerning exactly what constituted an "observation" or "measurement". Those issues were resolved by shifting from an ontological perspective, which tries to describe what really exists objectively "out there", to a practical or pragmatic perspective, which regards a physical theory as a useful collective conceptual human endeavor that aims to provide us with reliable expectations about our future experiences, for each of the alternative possible courses of action between which we are (seemingly) free to choose. As a collective endeavor, and in that sense as an objective theory, quantum mechanics is built on descriptions that allow us to communicate to others what we have done and what we have learned. Heisenberg strongly emphasized that this change in perspective converts quantum mechanics, in a very real sense, into a theory about "our knowledge": the relationships between experiential elements in our streams of consciousness become the core realities of a conceptual construction that aims to allow us to form, on the basis of what we already know, useful expectations about our future experiences, under the various alternative possible conditions between which we seem able to freely choose.

The paradoxical aspect of claiming the "physical state of a system" to be a representation of "our knowledge" is starkly exhibited by "Schrödinger's cat", whose quantum state is, according to this pragmatic approach, not determined until someone looks. Bohr escapes this dilemma by saying that the current quantum principles are insufficient to cover biological matter, but that approach leaves quantum mechanics fundamentally incomplete, and, in particular, inapplicable to the physical processes occurring in our brains.

In an effort to do better, von Neumann[3] showed how to preserve the rules and precepts of quantum mechanics all the way up to the interface with "experience", thereby preserving the general character of quantum mechanics as a theory that aims to provide reliable expectations about future experiences on the basis of present knowledge. Von Neumann's work brings into sharp focus the central problem of interest here, which is the connection between the properties specified in the quantum mechanical description of a person's brain and the experiential realities that populate that person's stream of consciousness. Bohr was undoubtedly right in saying that the orthodox precepts would be insufficient to cover this case. Additional ideas are needed, and the purpose of this article is to provide them.

The switch from classical mechanics to quantum mechanics preserves the idea that a physical system has a physically describable state. But the character of that state is changed drastically. Previously the physical state was conceived to have a well-defined meaning independently of any "observation". Now the physically described state has essentially the character of a "potentia" (an "objective tendency") for the occurrence of each one of a continuum of alternative possible "events". Each of these alternative possible events has both an experientially described aspect and also a physically described aspect: each possible "event" is a psychophysical happening. The experientially described aspect of an event is an element in a person's stream of consciousness, and the physically described aspect is a *reduction* of the set of objective tendencies represented by the prior state of that person's body/brain to the *part* of that prior state that is compatible with the increased knowledge supplied by the new element in that person's stream of consciousness. Thus the changing psychologically described state of that person's knowledge is correlated to the changing physically described state of the person's body/brain, and the changing physically described state entails, via the fundamental quantum probability formula, a changing set of weighted possibilities for future psychophysical events.

The practical usefulness of quantum theory flows from this lawful connection between a person's increasing knowledge and the changing physical state of his body/brain. The latter is linked to the surrounding physical world by the dynamical laws of quantum physics. This linkage allows a person to "observe" the world about him by means of the lawful relationship between the events in his stream of conscious experiences and the changing state of his body/brain.

It is worth noting that the physically described aspect of the theory has lost its character of being a "substance", both in the philosophical sense that it is no longer *self-sufficient*, being intrinsically and dynamically linked to the mental, and also in the colloquial sense of no longer being *material*. It is *stripped of materiality* by its character of being merely a potentiality or possibility for a future event. This shift in its basic character renders the physical aspect somewhat idealike, even though it is conceived to represent objectively real tendencies.

The key "utility" property of the theory—namely the property of being useful—makes no sense, of course, unless we have, in some sense, some freedom to choose. An examination of the structure of quantum mechanics reveals that the theory has both a logical place for, and a logical need for, choices that are made in practice by the human actor/observers, but that are *not determined by the quantum physical state of the entire world, or by any part of it*. Bohr calls this choice "the free choice of experimental arrangement for which the quantum mechanical formalism offers the appropriate latitude."[4] This "free" choice plays a fundamental role in von Neumann's rigorous formulation of quantum mechanics, and he gives the physical aspect of this probing action the name "Process I".[5] This Process I action is not determined, even statistically, by the physically described aspects of the theory.

The fact that this choice made by the human observer/agent is not determined by the physical state of the universe means that *the principle of the causal closure of the physical domain is not maintained in contemporary basic physical theory*. It means also that Kim's formulation of *mind–body supervenience is not entailed by contemporary physical theory*. That formulation asserts that "what happens in our mental life is wholly dependent on, and determined by, what happens with our bodily processes."[6] Kim indicates that supervenience is a common element of all *physicalist theories*. But since supervenience is not required by basic (i.e., quantum) physics, the easy way out of the difficulties that have been plaguing physicalists for half a century, and that continue to do so, is simply to recognize that the precepts of classical physics, which are the scientific source of the notions of the causal closure of the physical, and of supervenience, do not hold in real brains, whose activities are influenced heavily by quantum processes.

Before turning to the details of the quantum mechanical treatment of the relationship between mind and brain, I shall make a few comments on Kim's attempted resolution of the difficulties confronting the classical physicalist approach. The essential problem is the mind–body problem. Kim divides this problem into two parts: the problem of mental causation and the problem of consciousness. The problem of mental causation is: "How can the mind exert its causal powers in a world that is fundamentally physical?" The problem of consciousness is: "How can a thing such as consciousness exist in a physical world, a world consisting ultimately of nothing but bits of matter distributed over spacetime in accordance with the laws of physics."

From a modern physics perspective, the way to resolve these problems is immediately obvious: simply recognize that the assumption that the laws of physics pertain to "bits of matter distributed over spacetime in accordance with the laws of physics" is false. Indeed, that idea has, for most of the twentieth century, been asserted by orthodox physicists to be false, along with the assumption that the world is physical in the classical sense. Quantum mechanics builds upon the obvious real existence of our streams of conscious experiences, and provides also, as we shall see, a natural explanation of their causal power to influence physical properties. Thus the difficulties that have beset physicalists for five decades, and have led to incessant controversies and reformulations, stem, according to the perspective achieved by twentieth-century physics, directly from the fact that the physicalist assumptions not only do not follow from basic precepts of physics, but instead, directly contradict them. The premises of classical physicalists have been, from the outset, incredibly out of step with the physics of their day.

Kim tries at one point to squash the notion that the difficulties with physicalism can be avoided by accepting some form of dualism. But the dualism that he considers is a Cartesian dualism, populated with mysterious souls. However, quantum mechanics is science! The experientially described realities that occur in quantum theory are the core realities of science. They are the ideas that we are able to communicate to others pertaining to what we have done and what we have learned. These descriptions are essentially descriptions of (parts of) the accessible contents of the streams of consciousness of real living observer-agents. Criticizing dualism in the form advanced by Descartes during the seventeenth century instead of in the form employed in contemporary science is an indication that philosophers of mind have isolated themselves in a hermetically sealed world, created by considering only what other philosophers of mind have said, or are saying, with no opening to the breezes that bring word of the highly pertinent revolutionary change that had occurred in basic science decades earlier.

Kim's main argument leads to the conclusion that a physicalist must, for each conscious experience, choose between two options: either that experience is causally powerless, or it must be defined to be the causally efficacious brain activity that possesses its causal power. Kim himself admits that neither option is very palatable. The idea that our beliefs, desires, and perceptions, including our pains, have no effects upon our actions is regarded by Kim as unacceptable. Thus he opts for what he claims to be the only alternative available to the rational physicalist, namely that each such efficacious experience must be (defined to be) a causally efficacious brain activity that causes its effects: "If anything is to exercise causal power in the physical domain it must be an element in the physical domain or be reducible to it."[7] "Only physically reducible mental properties can be causally efficacious."[8]

That a conscious experience can be defined to be a physical activity, described in the mathematical language of physics, is certainly a hard pill to swallow. Fortunately, it is not true in quantum theory, where the physically described state represents merely an "objective tendency" for a psychophysical event to occur. However, the mind–brain identity that Kim describes does have a less-problematic analog in quantum theory. Each actual event has two sides: an experience; and a reduction of the prior state of the body/brain to one that incorporates into the physically described world a causal aspect conceptually represented in the intentional aspect of the experience. This is the essential core of the orthodox von Neumann/Heisenberg quantum position. It will be elaborated upon here.

Kim's solution has another apparent defect: different aspects of a person's apparently highly integrated stream of consciousness have fundamentally different statuses, in regard to their connections to that person's brain. Beliefs, desires, and percepts are defined to be brain activities, whereas colors and other "qualia" are not brain activities and are not causally efficacious. But how can your desire for a beautiful painting be simply a brain activity, whereas the particular colors that combine to excite this desire are epiphenomenal qualities having no effects on your brain?

The physicalist assumption has apparently led, after 50 years of development, to conclusions that are far from ideal. These conclusions fail to explain either why our conscious experiences should exist at all in a world that is dynamically and logically complete at the physical level of description, or how they can *be* physical properties that do not entail the existence of the experiential "feel" that characterize them. These long-standing difficulties arise directly from accepting the classical conception of the nature and properties of the physically described aspects of our description of the world. They are resolved in a natural way by accepting the quantum mechanical conception of the nature and properties of the aspects of the world that are described in physical terms: i.e., in terms of properties specified by assigning mathematically properties to spacetime regions.

In the following sections I shall explain how these difficulties are resolved by accepting the quantum conception of the physical.

13.3 Quantum Mechanics: The Rules of the Game

13.3.1 The Basic Formula

Quantum mechanics is a superstructure erected upon a basic formula. This formula specifies the probability that a probing action that is describable in everyday language, refined by the concepts of classical physical theory, will produce a *pre-specified* possible experienced outcome that is described in the same kind of terms. First a *preparing* action must be performed. Its outcome is represented by a (quantum) state of the prepared system. Then a probing action is chosen and performed. The elementary probing actions are actions that either produce a pre-specified outcome "Yes", or fail to produce that pre-specified outcome.

To achieve generality I shall adopt the density matrix formulation described by von Neumann. In this formulation the physical state of a system is represented by a matrix that is called the density matrix. It is traditionally represented by the symbol ρ. A measurement or observation on such a system is effected by means of a probing action, which is represented by a matrix, traditionally designated by the symbol P, or by a P with a subscript, that satisfies $PP = P$. Such a matrix/operator is called a *projection operator*. The quantum game is like "twenty questions": the observer-agent "freely poses" a question with an observable answer "Yes" or "No". This question, and the probing action corresponding to it, are represented in the formalism by some projection operator P. Nature then returns an answer "Yes" or "No". The probability that the answer is "Yes" is given by the basic probability equation of quantum mechanics:

$$\langle P \rangle = \frac{\operatorname{tr} P\rho}{\operatorname{tr} \rho}.$$

In order not to lose nonphysicists, but rather to get them into the quantum swing of things, and allow them to play this wonderful game, I shall spell out what this equation means in the simple case in which the matrices involved have just two rows and two columns. In this case each matrix/operator has four elements, which are specified by the four numbers $\langle 1|M|1\rangle$, $\langle 1|M|2\rangle$, $\langle 2|M|1\rangle$, and $\langle 2|M|2\rangle$. The index on the left specifies the horizontal row, and the index on the right specifies the vertical column of the matrix in which the matrix element is to be placed. The rule of matrix multiplication says, for any two matrices M and N, and any pair of two-valued indices i and j,

$$\langle i|MN|j\rangle = \langle i|M|k\rangle\langle k|N|j\rangle,$$

where one is supposed to sum over the two possible values of the repeated index k. For any M,

$$\mathrm{tr}\, M = \langle k|M|k\rangle,$$

where one is again supposed to sum over the (two in this case) different possible values (1 and 2) of the index k.

This case of a system represented by two-by-two matrices is physically very important: it covers the case of the "spin" degree of freedom of an electron. Once one sees how quantum mechanics works in this simplest case, the generalization to all other cases is basically pretty obvious. So in order to keep nonphysicists on board I will spend a little time spelling things out in detail for this simple case.

Pauli introduced for this two-by-two case four particular matrices defined by

$$\langle 1|\sigma_0|1\rangle = 1, \langle 1|\sigma_0|1\rangle = 1,$$

$$\langle 1|\sigma_1|2\rangle = 1, \langle 2|\sigma_1|1\rangle = 1,$$

$$\langle 1|\sigma_2|2\rangle = -i, \langle 2|\sigma_2|1\rangle = i,$$

$$\langle 1|\sigma_3|1\rangle = 1, \langle 2|\sigma_3|2\rangle = -1,$$

with all other elements zero (i is the imaginary unit).

They satisfy $\sigma_j\sigma_j = \sigma_0 = I$, for all j, where I is the identity matrix; $\sigma_1\sigma_2 = i\sigma_3 = -\sigma_2\sigma_1$; $\sigma_2\sigma_3 = i\sigma_1 = -\sigma_3\sigma_2$; and $\sigma_3\sigma_1 = i\sigma_2 = -\sigma_1\sigma_3$. Most calculations can be done using just these products of the Pauli matrices.

For actions that probe the direction of the spin of the electron, the projection operator $P = \frac{1}{2}(I + \sigma_3)$ represents the probing action that corresponds to the query "Does the spin of the electron point in the direction of the axis number 3?" It is also the density matrix that represents the spin state of the electron if the answer to that query is "Yes". In the higher-dimensional cases, if ρ is the density matrix prior to the probing action, then the density matrix after a probing action P that produces the answer "Yes" is $P\rho P$ (up to a possible positive multiplicative factor that drops out of the probability formula). If the feedback is "No", then ρ is reduced to $P'\rho P'$, with $P' = (1 - P)$.

Suppose one has prepared the spin state of the electron by performing the probing action corresponding to $P = \frac{1}{2}(I + \sigma_3)$ and has received the answer "Yes". This means that the density matrix for the system is now (known to be the state represented by) $\rho = \frac{1}{2}(I + \sigma_3)$. Suppose one now performs the probing action corresponding to the query "Does the spin point in the direction of axis number 1?" The corresponding P is $\frac{1}{2}(I + \sigma_1)$. Thus the probability that the answer is "Yes" is

$$\frac{\text{tr} \, \frac{1}{2}(I + \sigma_1)\frac{1}{2}(I + \sigma_3)}{\text{tr} \, \frac{1}{2}(I + \sigma_3)} = \frac{1}{2}.$$

This simplest example beautifully epitomizes the general case. It illustrates very accurately how the basic probability formula is used in actual practice.

The basic probability formula and its workings constitute the foundation of the quantum mechanical conception of the connection between the aspects of our scientific understanding of nature described in the language that we use to describe the pertinent perceptual and felt contents of our streams of conscious experiences and the aspects described in the mathematical language of physics.

13.3.2 Classical Description

Bohr wrote:

> ... we must recognize above all that, even when phenomena transcend the scope of classical physical theories, the account of the experimental arrangement and the recording of observations must be given in plain language, suitably supplemented by technical physical terminology. This is a clear logical demand, since the very word "experiment" refers to a situation where we can tell others what we have done and what we have learned.[9]

and

> ... it is imperative to realize that in every account of physical experience one must describe both experimental conditions and observations by the same means of communication as the one used in classical physics.[10]

This demand that we *must use* the known-to-be-fundamentally-false concepts of classical physical theories as a fundamental part of quantum mechanics has often been cited as the logical incongruity that lies at the root of the difficulties in arriving at a rationally coherent understanding of quantum mechanics: of an understanding that goes beyond merely understanding how to use it in practice. So I focus next on the problem of reconciling the quantum and classical concepts, within the context of a theory of the mind–brain connection.

13.3.3 Quasi-classical States of the Electromagnetic Field

There is one part of quantum theory in which a particularly tight and beautiful connection is maintained between classical mechanics and quantum mechanics. This is the simple harmonic oscillator (SHO). With a proper choice of units, the energy (or Hamiltonian) of the system has the simple quadratic form $E = H = \frac{1}{2}(p^2 + q^2)$, where q and p are the coordinate and momentum *variables* in the classical case, and are the corresponding *operators* in the quantum case. In the classical case the trajectory of the "particle" is a circle in q–p space of radius $r = (2E)^{\frac{1}{2}}$. The angular velocity is constant and independent of E, and in these special units is $\omega = 1$: one radian per unit of time. The lowest-energy classical state is represented by a point at rest at the "origin" $q = p = 0$.

The lowest-energy quantum state is the state—i.e., projection operator P—corresponding to a Gaussian wave function that in coordinate space is $\Psi(q) = Ce^{-\frac{1}{2}q^2}$ and in momentum space is $\Psi(q) = Ce^{-\frac{1}{2}p^2}$, where C is $2^{\frac{1}{4}}$. If this ground state is shifted in q–p space by a displacement (Q, P), one obtains a state—i.e., a projection operator P—labeled by $[Q, P]$ which has the following important property: if one allows this quantum state to evolve in accordance with the quantum mechanical equations of motion, then it will evolve into the set of states labeled by $[Q(t), P(t)]$, where the (center) point $(Q(t), P(t))$ moves on a circular trajectory that is identical to the one followed by the classical point particle.

If one puts a macroscopic amount of energy E into this quantum state, then it becomes "essentially the same as" the corresponding classical state. Thus if the energy E in this one degree of freedom is the energy per degree of freedom at body temperature, then the quantum state, instead of being confined to an *exact point* $(Q(t), P(t))$ lying on a circle of (huge) radius $r = 10^6$ in q–p space, will be effectively confined, owing to the Gaussian fall-off of the wave functions, to a disc of unit radius centered at that point $(Q(t), P(t))$. Given two such states, $[Q, P]$ and $[Q', P']$, their overlap, defined by the trace of the product of these two projection operators, is $e^{\frac{1}{2}d^2}$, where d is the distance between their center points. On this 10^6 scale, the unit size of the quantum state becomes effectively zero. And if the energy of this classical SHO state is large on the thermal scale, then its motion, as defined by the time evolution of the projection operator $[Q(t), P(t)] \equiv P(P(t), Q(t))$, will be virtually independent of the effects of both environmental decoherence, which arises from subtle quantum-phase effects, and thermal noise, for reasons essentially the same as the reasons for the negligibility of these effects on the classically describable motion of the pendulum on a grandfather clock.

Notice that the quantum state $[Q, P]$ is completely specified by the corresponding classical state (Q, P): the quantum mechanical spreading around this point is not only very tiny on the classical scale; it is also completely fixed: the width of the Gaussian wave packet associated with our Hamiltonian is fixed, and independent of both the energy and phase of the SHO.

We are interested here in brain dynamics. Everyone admits that at the most basic dynamical level the brain must be treated as a quantum system: the classical laws fail at the atomic level. This dynamics rests upon myriads of microscopic processes, including flows of ions into nerve terminals. These atomic-scale processes must in principle be treated quantum mechanically. But the effect of accepting the quantum description at the microscopic level is to inject quantum uncertainties/indeterminacies at this level. Yet introducing even small uncertainties/indeterminacies at microscopic levels into these nonlinear systems possessing lots of releasable stored chemical energy has a strong tendency — the butterfly effect — to produce very large macroscopic effects later on. Massive parallel processing at various stages may have a tendency to reduce these indeterminacies, but it is pure wishful thinking to believe that these indeterminacies can be completely eliminated in all cases, thereby producing brains that are completely deterministic at the macroscopic level. *Some* of the microscopic quantum indeterminacy *must* at least occasionally make its way up to the macroscopic level.

According to the precepts of orthodox quantum mechanics, these macroscopic quantum uncertainties are resolved by means of Process I interventions, *whose forms are not specified by the quantum state of the universe, or any part thereof*. What happens in actual practice is determined by conscious choices "for which the quantum mechanical formalism offers the appropriate latitude". No way has yet been discovered by quantum theorists to circumvent this need for some sort of intervention that is not determined by the orthodox physical laws of quantum physics. In particular, environmental decoherence effects certainly do not, by themselves, resolve this problem of reconciling the quantum indeterminacy, which irrepressibly bubbles up from the microscopic levels of brain dynamics, with the essentially classical character of our descriptions of our experiences of "what we have done and what we have learned".

The huge importance of the existence and properties of the quasi-classical quantum states of SHOs is this: if the projection operators P associated with our experiences are projection operators of the kind that instantiate these quasi-classical states, then we can rationally reconcile the demand that the dynamics of our brains be fundamentally quantum mechanical with the demand that our descriptions of our experiences of "what we have done and what we have learned" be essentially classical. This arrangement would be a natural upshot of the fact that our experiences correspond to the actualization of strictly quantum states that are both specified by classical states, and also closely mimic the properties of their classical counterparts, *apart from the fact that they represent only potentialities*, and hence will be subject, just like Schrödinger's macroscopic cat, to the actions of the projection operators associated with our probing actions. This quantum aspect entails that, by virtue of the quantum Zeno effect, which follows from the basic quantum formula that connects our conceptually described observations to physically described quantum jumps, we can understand *dynamically* how our conscious choices can affect our subsequent thoughts and actions: we can rationally explain, by using the basic principles of orthodox contemporary physics, the causal efficacy of our conscious thoughts in the physical world, and thereby dissolve the physicalists' dilemma.

I shall now describe in more detail how this works.

13.4 The Mind–Brain Connection

The general features of this quantum approach to the mind–brain problem have been described in several prior publications.[2,11] In this section I will present a specific model based on the general ideas described in those publications.

Mounting empirical evidence[12] suggests that our conscious experiences are connected to brain states in which measurable components of the electromagnetic field located in spatially well-separated parts of the brain are oscillating with the same frequency, and in phase synchronization. The model being proposed here assumes, accordingly, that the brain correlate of each conscious experience is an EM (electromagnetic) excitation of this kind. More specifically, each Process I probing action is represented quantum mechanically in terms of a projection operator that is the quasi-classical counterpart of such an oscillating component of a classical EM field.

The central idea of this quantum approach to the mind–brain problem is that each Process I intervention is the physical aspect of a psychophysical event whose psychologically described aspect is the conscious experience of intending to do, or choosing to do, some physical or mental action. The physical aspect of the "Yes" answer to this probing event is the actualization, by means of a quantum reduction event, of a pattern of brain activity called a "template for action". A *template for action* for some action X is a pattern of physical (brain) activity which if held in place for a sufficiently long time will tend to cause the action X to occur. The psychophysical linkage between the conscious intent and the linked template for action is supposed to be established by trial-and-error learning.

A prerequisite for trial-and-error learning of this kind is that mental effort be causally efficacious in the physically described world. Only if conscious choices and efforts have consequences in the physically described world can an appropriate correlation connecting the two be mechanically established by trial-and-error learning. With no such connection, the conscious intention could become completely opposed to the correlated physical action, with no way to activate a corrective physical measure.

The feature of quantum mechanics that allows a person's conscious choices to influence that person's physically described brain process in the needed way is the so-called "quantum Zeno effect". This quantum effect entails that if a sequence of very similar Process I probing actions occur in sufficiently rapid succession, then the affected component of the physical state will be forced, with high probability, to be, at the particular sequence of times t_i at which the probing actions are made, exactly the sequence of states specified by the sequence of projection operators $P_h(t_i)$ that specify the "Yes" outcomes of the sequence of Process I actions. That is, the affected component of the brain state—for example some template for action—will be forced, with high probability, *to evolve in lock step with a sequence of "Yes" outcomes of a sequence of "freely chosen" Process I actions, where "freely chosen" means that these Process I actions are not determined, via any known law, by the physically described state of the universe!* This coercion of a physically described aspect of a brain process to evolve in lock step with the "Yes" answers to a sequence of Process I probing actions that are free of any known physically described coercion, but that seem to us to be freely chosen by our mental processes, is what will presently be demonstrated. It allows physically uncoerced conscious choices to affect a physically described process that will, by virtue of the basic probability formula, have experiential consequences.

The repetition rate (attention density!) in the sequence of Process I actions is assumed to be controlled by conscious effort. In particular, in the model being described here, where the projection operators $P(t_i)$ are projection operators $[Q(t_i), P(t_i)]$ that are quasi-classical states of SHOs, the size of the intervals $(t_{i+1} - t_i)$—being a feature of the sequence of "freely chosen" Process I probing actions—is taken to be under the immediate control of the psychological aspect of the probing action.

I describe the quantum properties of the EM field in the formulation of relativistic quantum field theory developed by Tomonaga[13] and Schwinger[14], which generalizes the idea of the Schrödinger equation to the case of the electromagnetic field. One can imagine space to be cut up into very tiny regions, in each of which the values of the six numbers that define the electric and magnetic fields in that region are defined. In case the field in that region is executing simple harmonic oscillations, we can imagine that each of the six values is moving in a potential well that produces the motion of an SHO. If the Process I action is specified by a "Yes" state that is a coordinated synchronous oscillation of the EM field in many regions, $\{R_1, R_2, R_3, \ldots\}$, then this state, if represented quantum mechanically, consists of some quasi-classical state $[Q_1, P_1]$ in R_1, *and* some quasi-classical state $[Q_2, P_2]$ in R_2, *and* some quasi-classical $[Q_3, P_3]$ in R_3, etc. The state P of this combination is the *product* of these $[Q_i, P_i]$s, each of which acts in its own SHO space, and acts like the unit operator (i.e., unity or "one") in all the other spaces. This product of P_ns, all evaluated at time t_i, is the $P_h(t_i)$ that is the brain aspect of the "Yes" answer to the Process I query that occurs at time t_i. The *quantum* frequency of the state represented by this $P_h(t_i)$ is the sum of the *quantum* frequencies of the individual regions, and is the total number of quanta in the full set of SHOs. However, the period of the periodic motion of the classical EM field remains 2π, in the chosen units, independently of how many regions are involved, or how highly excited the states of the SHOs in the various regions become. This smaller frequency is the only one that the classical state knows about: it is the frequency that characterizes the features of brain dynamics observed in EEG and MEG measurements.

The sequence of $P_h(t_i)$s that is honed into the observer/agent's structure by trial-and-error learning is a sequence of $P_h(t_i)$s that occurs when the SHO template for action is held in place by effort. Learning is achieved by effort, which increases attention density, and holds the template for action in place. Thus if H_0 is the Hamiltonian that maintains this SHO motion, then for the honed sequence,

$$P_h(t_{i+1}) = e^{-iH_0(t_{i+1}-t_i)} P_h(t_i) e^{iH_0(t_{i+1}-t_i)}$$

But in the new situation there may be disturbing physical influences that tend to cause a deviation from the learned SHO motion. Suppose that on the time scale of $(t_{i+1} - t_i)$ the disturbance is small, so that the perturbed evolution starting from $P_h(t_i)$ can be expressed in the form

$$P(t_{i+1}) = e^{-iH_i(t_{i+1}-t_i)} e^{-iH_0(t_{i+1}-t_i)} P_h(t_i) \, e^{iH_0(t_{i+1}-t_i)} e^{iH_i(t_{i+1}-t_i)}$$
$$= e^{-iH_i(t_{i+1}-t_i)} P_h(t_{i+1}) \, e^{iH_i(t_{i+1}-t_i)},$$

where H_i is bounded.

According to the basic probability formula, the probability that this state $P(t_{i+1})$ will be found, if measured/observed, to be in the state $P_h(t_{i+1})$ at time t_{i+1} is (using tr $P_h(t_i) = 1$)

$$\text{tr } P_h(t_{i+1}) \, e^{-iH_i(t_{i+1}-t_i)} P_h(t_{i+1}) \, e^{iH_i(t_{i+1}-t_i)}.$$

Inserting the leading and first-order terms $(1 \pm iH_i(t_{i+1} - t_i))$ in the power series expansion of $e^{\pm iH_i(t_{i+1}-t_i)}$ and using $PP = P$, and the fact that tr $AB =$ tr BA, for all A and B, one finds that the term linear in $(t_{i+1} - t_i)$ vanishes identically.

The vanishing of the term linear in $(t_{i+1} - t_i)$ is the basis of the quantum Zeno effect. If one considers some finite time interval and divides it into small intervals $(t_{i+1} - t_i)$ and looks at a product of factors $(1 + c(t_{i+1} - t_i)^n)$, then if n is bigger than one, the product will tend to unity (one) as the size of the intervals $(t_{i+1} - t_i)$ tend to zero. But this means that the basic probability formula of quantum mechanics requires that, as the step sizes $(t_{i+1} - t_i)$ tend to zero, the evolving state of the system being probed by the sequence of probing actions will have a probability that tends to one (unity) *to evolve in lock step with the set of "Yes" answers to the sequence of probing actions*, provided the initial answer was "Yes". But the forms of the projection operators $P_h(t_i)$ and the timings of the probing actions are not determined by the laws of orthodox quantum theory: they are "freely chosen". Hence orthodox quantum theory accommodates in a natural way the capacity of a person's conscious intentional choices to influence the processes occurring in his or her physically described brain, and to influence them in a way that will tend to produce intended consequences.

The point of this derivation is that it is expressed in terms of brain states that are macroscopic, and that correspond to classically describable states of the electromagnetic field measured by EEG and MEG procedures. Even though these states contain huge amounts of energy, nevertheless, if we accept the principle that the underlying brain dynamics must in principle be treated quantum mechanically, and, accordingly, replace these classical states by their quasi-classical counterparts, which represent potentialities that are related to experience only via the basic equation, then the principles of orthodox von Neumann quantum mechanics provide a rationally coherent way of understanding the mind–brain connection in a way that escapes the horns of the physicalists' dilemma: it gives each person's intentional conscious choices the power to causally affect the course of events in his or her quantum mechanically described brain, and to influence it in a way that serves these intentions.

References

1 J. Kim, *Physicalism, or Something Near Enough* (Princeton University Press, Princeton NJ, 2005).
2 H. P. Stapp, *Mind, Matter, and Quantum Mechanics*, second edition (Springer, Berlin, New York, 2004). H. P. Stapp, *Mindful Universe: Quantum Mechanics and the Participating Observer* (Springer, Berlin, New York, 2007). H. P. Stapp, Quantum Interactive Dualism: An Alternative to Materialism, *J. Consc. Stud.* **12**, no. 11, 43–58 (2005). J. M. Schwartz, H. P. Stapp, and M. Beauregard, Quantum Theory in Neuroscience and Psychology: A Neurophysical Model of the Mind/Brain Interaction, *Phil. Trans. Roy. Soc. B*, **360** (1458) 1306 (2005).
3 J. von Neumann, *Mathematical Foundations of Quantum Mechanics* (Princeton University Press, Princeton NJ, 1955), translated from the 1932 German original by R. T. Beyer.
4 N. Bohr, *Atomic Physics and Human Knowledge* (Wiley, New York, 1958), p. 73.
5 Ref. 3, pp. 351, 418, 421.
6 Ref. 1, p. 14.
7 Ref. 1, pp. 170–171.
8 Ref. 1, p. 174.
9 Ref. 4, p. 72.
10 Ref. 4, p. 88.
11 H. P. Stapp in *Physics and Whitehead: Quantum, Process, and Experience*, edited by T. Eastman and H. Keeton (SUNY Press, Albany NY, 2004). H. P. Stapp, Light as Foundation of Being, in *Quantum Implications: Essays in Honor of David Bohm* (Routledge and Kegan Paul, London & New York, 1987). H. P. Stapp, On the Unification of Quantum Theory and Classical Physics, in *Symposium on the Foundations of Modern Physics: 50 years of the Einstein–Podolsky–Rosen Gedankenexperiment*, edited by P. Lahti and P. Mittelstaedt (World Scientific, Singapore, 1985). J. R. Klauder and E. C. G. Sudarshan, *Fundamentals of Quantum Optics* (Benjamin, New York, 1968). The quasi-classical states used in the present paper are the projection operators corresponding to the "coherent states" described in this reference.
12 J. Fell, G. Fernandez, P. Klaver, C. Elger, and P. Fries, Is Synchronized Neuronal Gamma Activity Relevant for Selective Attention?, *Brain Res. Rev.* **42**, 265–272 (2003). A. Engel, P. Fries, and W. Singer, Dynamic Predictions: Oscillations and Synchrony in Top–Down Processing, *Nat. Rev. Neurosci.* **2**, 704–716 (2001).
13 S. Tomonaga, On a Relativistically Invariant Formulation of the Quantum Theory of Wave Fields, *Prog. Theor. Phys.*, **1**, 27–42 (1946).
14 J. Schwinger, Theory of Quantized Fields I, *Phys. Rev.*, **82**, 914–927 (1951).

14 A Model of the Quantum–Classical and Mind–Brain Connections, and the Role of the Quantum Zeno Effect in the Physical Implementation of Conscious Intent

14.1 Introduction

The basic problem in the interpretation of quantum mechanics is to reconcile the fact that our observations are describable in terms of the concepts of classical (i.e., nineteenth-century) physics, whereas the atoms from which our measuring devices and our physical body/brains are made obey the laws of quantum (twentieth-century) physics. The direct application of the microscopic atomic laws to macroscopic aggregates of atoms is well defined, but the thus-defined aggregates of atoms are not describable in classical terms.

The basic problem of the philosophy of mind, and indeed of all philosophy, is to understand the connection of our conscious thoughts to the physically described world. No feature, configuration, or activity of the physical world, as it is conceived of and described in classical physics, *is* the experiential quality that characterizes our conscious thoughts, ideas, and feelings. Something beyond the classically conceived physical world seems to be needed in the full inventory of what exists.

The obvious solution to this second problem is to recognize that the basic precepts of classical physics were replaced during the first part of the twentieth century by those of quantum theory. This improved physical theory brings conscious human observer/agents into physics in an essential way that renders the classical conceptions of our bodies, including our brains, fundamentally deficient. The new theory accommodates a mechanism that allows our conscious thoughts to influence our bodily actions without being reducible to any physically describable feature or activity. Hence accepting the precepts of contemporary physics provides an adequate and suitable basis of a rational answer to the second question. But it leaves untouched the first question, about the basis within a quantum universe of the classical describability of our perceptions.

Mounting neuroscientific evidence indicates that our conscious intentions are closely linked to synchronous ~40 Hz oscillations of the electromagnetic field at many well-separated brain sites.[1] This result points to the importance of classically describable simple harmonic oscillator (SHO) motions in the description and understanding of our conscious intentions and their physical effects. But a focus on SHO motions opens the door to a relatively simple solution of the problem of the connection between the classical character of our descriptions of our perceptions and the quantum character of our description of the physical dynamics. The so-called "coherent states" associated with SHO motions connect quantum concepts to classical concepts in just the way needed to achieve a simple, rational, simultaneous solution of the problems of the quantum–classical and mind–brain connections. The purpose of this chapter is to describe an exactly solvable model that exhibits in a clear way the basic elements of this resolution of these two problems.

In this model the causal effectiveness of our conscious intentions rests heavily upon the quantum Zeno effect. This is a strictly quantum mechanical effect that has been advanced elsewhere[2] as the dynamical feature that permits "free choices" on the part of an observer to influence his or her bodily behavior.

The intervention by the observer into physical brain dynamics is an essential feature of orthodox (von Neumann) quantum mechanics. Within the von Neumann quantum dynamical framework, this intervention can, with the aid of the quantum Zeno effect, cause a person's brain to behave in a way that causes the body to act in accord with the person's conscious intent. However, previous accounts of this mechanism, although strictly based on the mathematical principles of quantum mechanics, have been directed primarily at neuroscientists and philosophers, and have therefore been largely stripped of equations. The present model is so simple that the equations and their meanings can be presented in a way that should be understandable both to physicists and to sufficiently interested nonphysicists who are not troubled by simple equations.

On the other hand, the use of quantum mechanical effects in brain dynamics might seem problematic, because it depends on the existence of a macroscopic quantum effect in a warm, wet, noisy brain. It has been argued that *some* such effects will be destroyed by environmental decoherence[3] Those arguments do cover many macroscopic quantum mechanical effects, but they fail, for the reasons described below, to upset the quantum Zeno effect at work here.

In section 14.2 I shall review the well-known properties of a system of two coupled SHOs. In section 14.3 I shall use those results, and the closely related properties of the associated quantum "coherent states", to construct a mathematically solvable quantum mechanical model of the connection between conscious intent and brain activity. In section 14.4 I describe the conclusions to be drawn.

14.2 Coupled Oscillators in Classical Physics

It is becoming increasingly clear that at least some of our normal conscious experiences are associated with $\sim$40 Hz synchronous oscillations of the electromagnetic fields at a collection of brain sites.[4] These sites are evidently dynamically coupled. And the brain appears to be, in some sense, approximately described by classical physics. So I begin by recalling some elementary facts about coupled classical simple harmonic oscillators (SHOs).

In suitable units the Hamiltonian for two SHOs of the same frequency is

$$H_0 = \tfrac{1}{2}(p_1^2 + q_1^2 + p_2^2 + q_2^2).\tag{14.1}$$

If we introduce new variables via the canonical transformation

$$P_1 = \frac{1}{\sqrt{2}}(p_1 + q_2),\tag{14.2}$$

$$Q_1 = \frac{1}{\sqrt{2}}(q_1 - p_2),\tag{14.3}$$

$$P_2 = \frac{1}{\sqrt{2}}(p_2 + q_1),\tag{14.4}$$

$$Q_2 = \frac{1}{\sqrt{2}}(q_2 - p_1),\tag{14.5}$$

and replace the above H_0 by

$$H = (1 + e)\tfrac{1}{2}(P_1^2 + Q_1^2) + (1 - e)\tfrac{1}{2}(P_2^2 + Q_2^2),\tag{14.6}$$

then this H expressed in the original variables is

$$H = H_0 + e(p_1 q_2 - q_1 p_2).\tag{14.7}$$

If $e \ll 1$, then the term proportional to e acts as a weak coupling between the two SHOs whose motions for $e = 0$ would be specified by H_0.

The Poisson bracket (classical) equations of motion for the coupled system are, for any x,

$$\frac{dx}{dt} = \{x, H\} = \sum_j \left(\frac{\partial x}{\partial q_j}\frac{\partial H}{\partial p_j} - \frac{\partial x}{\partial p_j}\frac{\partial H}{\partial q_j} \right).\tag{14.8}$$

They give

$$\frac{dp_1}{dt} = -q_1 + p_2 e, \tag{14.9}$$

$$\frac{dp_2}{dt} = -q_2 - p_1 e, \tag{14.10}$$

$$\frac{dq_1}{dt} = p_1 + q_2 e, \tag{14.11}$$

$$\frac{dq_2}{dt} = p_2 - q_1 e. \tag{14.12}$$

A solution is

$$p_1 = \tfrac{1}{2}C[\cos(1+e)t + \cos(1-e)t]$$
$$= C \cos t \cos et, \tag{14.13}$$

$$q_2 = \tfrac{1}{2}C[\cos(1+e)t - \cos(1-e)t]$$
$$= -C \sin t \sin et, \tag{14.14}$$

$$p_2 = \tfrac{1}{2}C[-\sin(1+e)t + \sin(1-e)t]$$
$$= -C \cos t \sin et, \tag{14.15}$$

$$q_1 = \tfrac{1}{2}C[\sin(1+e)t + \sin(1-e)t]$$
$$= C \sin t \cos et. \tag{14.16}$$

The second line of each equation follows from the trigonometric formulas for sines and cosines of sums and differences of their argumants. A common phase ϕ can be added to the argument of every sine and cosine in the first line of each of the four equations. This leads to the addition of this phase to the argument t, but not the argument et, in the second line of each of the four equations.

These equations specify the evolving state of the two SHO systems by a trajectory in (p_1, q_1, p_2, q_2) space.

When we introduce the quantum corrections by quantizing this classical model, we obtain an almost identical quantum mechanical description of the dynamics. In a very well known way, the Hamiltonian H_0 goes over to (I use units where Planck's constant is 2π)

$$H_0 = \tfrac{1}{2}\left(p_1^2 + q_1^2 + p_2^2 + q_2^2\right)$$
$$= \left(a_1^\dagger a_1 + \tfrac{1}{2}\right) + \left(a_2^\dagger a_2 + \tfrac{1}{2}\right). \tag{14.17}$$

The connection between the classical and quantum descriptions of the state of the system is very simple: the point in (p_1, q_1, p_2, q_2) space that represents the classical state of the whole system is replaced by a "wave packet"

that, insofar as the interventions associated with observations can be neglected, is a smeared out (Gaussian) structure centered for all times exactly
on the point that specifies the classical state of the system. That is, the quantum mechanical representation of the state specifies a probability distribution
of the form e^{-d^2}, where d is the distance from a center (of-the-wave-packet)
point (p_1, q_1, p_2, q_2), which is, at all times, exactly the point (p_1, q_1, p_2, q_2)
that is the classical representation of the state.

According to quantum theory, the operator $a_i^\dagger a_i = N_i$ is the number
operator that gives the number of quanta of type i in the state.

Thus, in the absence of any observations, the classical and quantum
descriptions are almost identical: there is, in the quantum treatment, merely a
small smearing-out in (p, q) space, which is needed to satisfy the uncertainty
principle.

This correspondence persists when the coupling is included. The coupling term in the Hamiltonian is

$$H_1 = \tfrac{1}{2}e\left(p_1 q_2 - q_1 p_2 - p_2 q_1 + q_2 p_1\right)$$
$$= \tfrac{1}{2}ie\left(a_1^\dagger a_2 - a_1 a_2^\dagger - a_2^\dagger a_1 + a_2 a_1^\dagger\right). \tag{14.18}$$

The Heisenberg (commutator) equations of motion generated by the quadratic Hamiltonian $H = H_0 + H_1$ give the same equations as before, but now
with operators in place of numbers. Consequently, the centers of the wave
packets will follow the classical trajectories also in the $e > 0$ case. The
radius of the orbit is the square root of twice the energy, measured in the
units defined by the quanta of energy associated with frequency of the SHO.

14.3 Application

With these well-known results in hand, we can turn to their application. The
above mathematics shows, for SHOs, a near identity between the classical
and quantum treatments, insofar as there are no observations. But if observations occur, then the quantum dynamics prescribes certain associated
actions on the quantum state.

The essential point here is that quantum theory, in the von Neumann/
Heisenberg formulation, describes the dynamical connection between conscious observations and brain dynamics.* To apply this theory, the classically described brain must first be converted to its quantum form. By virtue

---

* Von Neumann[4] brought the mind–brain connection into the formulation
in a clear way, as an application of the orthodox quantum precept that each
increment in our classically describable knowledge is represented in the

of the relationships described in section 14.2, this conversion is direct when the classical state that is connected to consciousness is an SHO state. And if no observations occur, the classical and quantum descriptions are essentially the same: the tiny smearing-out of the classical point to the narrow Gaussian centered on the classical point is of negligible significance.

The observer, in order to get information about what is going on about him into his stream of consciousness, must, according to orthodox quantum mechanics, initiate probing actions. According to the development of the theory of von Neumann[6] described in references 7, the brain does most of the work. It creates, in an essentially mechanical way based on trial-and-error learning, and also upon the current quantum state of the brain, a query/question. Each possible query is associated with a psychological projection into the future that specifies the brain's computed "expectation" about what the feedback from the query will be.**

The physical manifestation of this query is called "Process I" by von Neumann. It is a key and necessary element of the quantum dynamics: it resolves ambiguities that are not resolved by the physical laws of quantum mechanics, and it ties the physical description expressed in terms of the quantum mathematics to our communicable descriptions of our perceptions. This Process I probing action is *not* the famous statistical element in quantum theory! It is needed both to specify what the statistical predictions will be *about*, and also to tie the abstract quantum mathematics to human perceptual experience, and hence to science.

In order to bring out the essential point, and also to embed the discussion comfortably into the common understandings of neuroscientists, who are accustomed to thinking that the brain is well described in terms of the concepts of classical physics, I shall consider first an approximation in which the brain is well described by classical ideas. Thus the two SHO

---

mathematical language of quantum mechanics by the action of associated projection operators on the prior state. Heisenberg[5] emphasized that if one wants to understand what is really happening, then the quantum state should be regarded as a "potentia" (objective tendency) for a real psychophysical event to occur.

** My idea here is to assume/postulate that if $\{P_1, P_2, P_3, \ldots\}$ is the set of Ps corresponding to all the questions that could be posed at time t, and $P(t)$ is the P_N that maximizes tr $P_N \rho(t)$, then the only question that could be asked at time t is $P(t)$. But whether this question will in fact be posed at time t could be influenced by experiential qualities. This would allow the *timings* of the probing actions to be determined in part by features of nature not represented in the physically described part.

states that we are focusing on are considered to be aspects of possible states of a classically described brain, which is also providing the potential wells in which these two SHOs move. It is the degrees of freedom of the brain associated with the first of these two SHOs that are, in the simple model being considered here, the possible brain correlates of the consciousness of the observer during the period of the experiment. Hence it is they that are affected by von Neumann's Process I. The second SHO, described by the pair of variables (p_2, q_2), represents environmental degrees of freedom. One sees from the second lines in each of the four equations (14.13–16) that in a period of duration $t = \pi(2e)^{-1}$ starting from time $t = 0$, the energy of the first SHO will, for $e \ll 1$, be fully transferred to the second SHO, provided no probing actions are made.

If no probing actions are made, then the conserved energy will oscillate with period $t = 2\pi(e)^{-1}$ back and forth between the two SHOs. Our interest here is in the effect upon this transfer of energy from the first SHO to the second SHO of a sufficiently rapid sequence of probing actions. What will be shown is that if the probing actions are sufficiently rapid on the scale of time $t = e^{-1}$, then the trajectory of (p_1, q_1) will tend to follow the uncoupled $(e = 0)$ trajectory.

The point, here, is that quantum mechanics has a built-in connection between a conscious intent and its physical effects. This connection is tied to the Process I probing actions, whose dynamical effects are specified by the quantum dynamical rules. Therefore our conscious intentions do not stand outside the dynamics as helpless, impotent witnesses, as they do in classical physics, but have *specified* dynamical effects. We are now in a position to examine what these effects are.

I assume that there is a rapid sequence of queries at a sequence of times $\{t_1, t_2, t_3, \ldots\}$. These queries will be based on expectations constructed by the brain on the basis of past experiences. The queries are represented in the quantum mathematics by a series of projection operators $\{P(t_1), P(t_2), P(t_3), \ldots\}$.* This sequence of projection operators represents a sequence of questions that ask whether the current state is on the "expected" track. This track is specified by the $e = 0$ trajectory, which represents expectations based on past experiences in which the holding-in-place effects of similar efforts have been present.

Up until now I have spoken as if the projection operators associated with the observations are projections onto a single quasi-classical state (i.e., onto one of the so-called "coherent" states). A projection upon such a state would involve fantastic precision. Each such state is effectively confined

---

* A *projection* operator P satisfies $PP = P$.

to a disc of unit size relative to an orbit radius C of about 10^6 in the units employed in equation (14.1).* However, it is possible (for our SHO case) to define more general operators that are projection operators (i.e., satisfy $PP = P$) apart from corrections of order, say, $< 10^{-3}$, by using the von Neumann lattice theorem.[9]

If one represents by $[P, Q]$ the projection operator that projects onto the Gaussian state centered at $(p, q) = (P, Q)$, then the lattice theorem says that the following identity holds:

$$\sum [mf, nf] = I, \qquad (14.19)$$

where $f = (2\pi)^{\frac{1}{2}}$, I is the identity operator, and the sum is over all integer values of m and n except $m = n = 0$. Moreover, the decomposition into different Gaussian components effected by this identity is unique. If one restricts the sum to the lattice points in a very large square region in (p_1, q_1) space, then the resulting operator P' is very nearly a projection operator.

For example, if the square region $S(C, 0)$ is centered at the SHO point $(C, 0)$ in the (p_1, q_1) space that we have been discussing, and has sides of length, say, one percent of the radius C of the unperturbed orbit, then each side of the square will be $10^4 f^{-1}$ units compared to the unit size associated with the Gaussian fall-off e^{-d^2}. In this case the associated quasi-projection operator $P' = P(C, 0)$ is essentially a projection operator onto the square region $S(C, 0)$ of (p_1, q_1) space: it will take any state vector, uniquely decomposed into the sum of terms specified in equation (14.19), approximately into the sub-sum over the terms occurring in P'.

Let $S(C \cos \phi, C \sin \phi)$ be the square, centered on $(C \cos \phi, C \sin \phi)$, obtained by rotating $S(C, 0)$ by ϕ, so that the line from its center point to the origin is parallel to two of its sides. The action of the unperturbed $(e = 0)$ Hamiltonian will take $S(C, 0)$ to $S(C \cos \phi, C \sin \phi)$ in time ϕ. It will also take $P(C, 0)$ to the quasi-projection operator $P(C \cos \phi, C \sin \phi)$ associated with the square $S(C \cos \phi, C \sin \phi)$. These results follow from the simple SHO dynamics in the unperturbed (decoupled) $e = 0$ case.

---

* This number 10^6 is roughly the square root of the thermodynamic energy per degree of freedom at body temperature, in energy units associated with equation (14.1), in which Planck's constant is 2π, and the angular velocity is one radian per unit of time. The unit of time in these units is about 4 ms for 40 Hz oscillations. An actual excited brain state should have energy significantly *greater* than thermal, but a higher energy makes our approximation even better.

The collapse rules of orthodox quantum dynamics are compactly stated in terms of the trace operation.*

The trace of the product of the "projection" operator $P\big(p_1(t, e), q_1(t, e)\big)$ centered on the perturbed orbit (where the two arguments are defined by equations (14.13) and (14.16)) with the "projection" operator $P\big(p_1(t, 0),$ $q_1(t, 0)\big) = P\big(C\cos t, C\sin t\big)$ centered on the unperturbed orbit is, to lowest order in t, $1 - \frac{1}{2}[(et)^2 100]$, where for a 40 Hz SHO the time unit is about 4 ms. The term $\frac{1}{2}[(et)^2 100]$ is the ratio of the displacement (of the perturbed square relative to the unperturbed square, namely $\frac{1}{2}(et)^2 C$), to the length of the side of the square, which is one percent of the radius C of the unperturbed orbit. The unperturbed square rotates rigidly with angular velocity unity, under the action of the unperturbed Hamiltonian, and the lowest-order $e > 0$ displacement is toward the origin $(p_1, q_1) = (0, 0)$. Consequently, the dynamics is essentially unchanged by rotations: the initial condition $(C, 0)$ plays no essential role.

According to the basic precepts of quantum theory, the (physical) "state" of the system at time t is specified by a "density matrix" (or "density operator"), usually denoted by $\rho(t)$. If the answer is "Yes", then the state immediately *after* the probing action at time t_i is $\rho(t_i+) = P(t_i)\rho(t_i-)P(t_i)$, where $\rho(t_i-)$ is the state immediately *before* the time t_i at which the question is posed. The operators $P(t_i)$ that occur on the right and left in $\rho(t_i+)$ project onto states that in our case are evolving at time t_i according to the unperturbed ($e = 0$) SHO motion. Hence for our case the first-order evolution forward in time from the probing time t_i is the same as the *unperturbed* ($e = 0$) evolution. This means that the small-time evolution forward in time by the time interval t from the time t_i of the ith probing action is given by the second lines of equations (14.13) and (14.16) with the arguments t in those two equations replaced by $t + t_i$ but the arguments et left unchanged.

The basic statistical law of quantum theory asserts that, *given the query specified by the projection operator $P(t)$, the probability that the answer will be "Yes" is* $\operatorname{tr}\rho(t+)$ divided by $\operatorname{tr}\rho(t-)$.

* The trace operation acting upon operators/matrices is defined by allowing the matrix (or operator) multiplication operation occurring in, say, $\operatorname{tr} AB$ to be extended cyclically, so that B acting to the right acts back on A. This means that for any pair of matrices/operators A and B, $\operatorname{tr} AB = \operatorname{tr} BA$. This property entails also that $\operatorname{tr} ABC = \operatorname{tr} BCA$. For any X, $\operatorname{tr} X$ is a number. In our case, $\operatorname{tr} P(P, Q)$ is essentially the area of the square $S(P, Q)$, measured in units of action given by Planck's constant, and $\operatorname{tr} P(P, Q)P(P', Q')$ is the area of the intersection of $S(P, Q)$ and $S(P', Q')$.

Note that the query, specified by $P(t)$ and by the time t at which $P(t)$ acts, must be specified *before* the statistical postulate can be applied!

If $\rho(t-)$ is, for the first probing time $t = t_1$, slowly varying over the square domain in (p_1, q_1) space, in the sense that $\mathrm{tr}\,[P, Q]\rho(t-)$ is essentially constant as (P, Q) varies over the square $S(C, 0)$, then the state immediately after the initial observation will be essentially the projection operator $P(C, 0)$ associated with that initial Process I probing action.

Under these conditions our equations show that for any (large) time T the density matrix $\rho(T)$ will be nearly equal to $P(C \cos T, C \sin T)$, provided the interval T is divided by observations into N equal intervals $t_{i+1} - t_i$, and $N(10eT)^2 N^{-2} \ll 1$. This condition entails both that *all* the answers will be "Yes" with probability close to unity, and also that the final $\rho(T)$ will be almost the same as the unperturbed "projection" operator $P(C \cos T, C \sin T)$.

Thus the rapid sequence of probing actions effectively holds the sequence of outcomes to the *expected* sequence. The affected brain states are constrained to follow the *expected* trajectory! This is the quantum Zeno effect, in this context.

This result means that if the probing actions come repetitiously at sufficiently short time intervals, then the probability that the state will remain on the unperturbed orbit for, say, a full second will remain high even though the perturbed $e > 0$ classical trajectory moves away from the unperturbed orbit by an amount of order C in time T of order e^{-1}.

The drastic slowing of the divergence of the actual orbit from the computed/expected orbit (circular in this case) is a manifestation of the quantum Zeno effect. The representation in the physically described brain of the probing action corresponding to the query "Is the brain correlate of the occurring percept the computed/expected state" is von Neumann's famous Process I, which lies at the mathematical core of von Neumann's quantum theory of the relationship between perception and brain dynamics.

14.4 Conclusions

The bottom line is that orthodox quantum mechanics has a built-in dynamical connection between conscious intent and its physically describable consequences. This connection fills a dynamical gap in the purely physically described quantum dynamical laws, and it allows certain specific mind–brain connections to be *deduced from the basic physics precepts relating mind and brain*. If a person can, by mental effort, *sufficiently increase the rate at which his Process I probing actions occur* [this is something not under the control, even statistically, of the physical laws of quantum mechanics], then that person can, by mental effort, quantum dynamically *cause* his brain/body to behave in a way that follows a pre-programmed trajectory, specified, say, by "expectations", instead of following the trajectory that it would follow if the von Neumann Process I probing actions do not occur in rapid succession. Because the causal origin of the Process I probing actions *is not specified, even statistically, by the presently known laws of physics,* there is in quantum mechanics a rational place for the experiential aspects of our description of nature to enter, irreducibly and efficaciously, into the determination of the course of certain physically described events.

I have focused here on the leading powers in t, in order to emphasize, and exhibit in a relatively simple way, the origin of the key result, which is that for small t on the scale, not of the exceedingly short period of the quantum mechanical oscillations, nor even on the $\sim$25 ms period of the $\sim$40 Hz scale of the classical oscillations, but on the much longer time scale of the *difference* of the periods of the two coupled modes, there will be, in this model, by virtue of the quantum mechanical effects associated with a rapid sequence of repeated probing actions, a strong tendency for the brain correlate of consciousness to follow the *expected* trajectory, in contrast to what would happen if only infrequent probing actions were made.

This analysis is based on a theory of the mind–brain connection that resolves in principle the basic interpretational problem of quantum theory, which is the problem of reconciling the classical character of our perceptions of the physical world with the non-classical character of the state of the world generated by the combination of the Schrödinger equation and the uncertainty principle. The theory resolves also the central problem of the philosophy of mind, which is to reconcile the apparent causal power of our conscious efforts with the laws and principles of physics. This relatively simple theory allows us to understand within the *dynamical framework* of orthodox (knowledge-associated-collapse) quantum physics the evident capacity of our conscious thoughts to influence our physical actions, and to become thereby integrated into the process of natural selection.

The discussion has focused so far on one very small region of the brain, or rather on one small region together with an environment into which it would, in the absence of probing observations, dissipate its energy. But the possible experiences of the relevant kind are associated with synchronous excitations in a large collection of such localized regions.[10] Following the principles of quantum *field* theory the associated quantum state is represented by a *tensor product* of states associated with the individual tiny regions.

There has been a lot of detailed theoretical work examining the effects of the fact that for a system that extends over an appreciable region of spacetime, the parts of the system located in different regions are coupled effectively to different degrees of freedom of the environment[11]. Insofar as these aspects of the environment are never observed, the predictions of quantum theory are correspondingly curtailed. In particular, *relative phases* of the wave function of the system associated with different regions become impossible to determine, and a "superposition" of spatially separated components becomes reduced to a "mixture".

In the model under consideration here, the components in different spacetime regions are different *factors* of a tensor product, rather than different terms of a superposition. In this case, the fact that different regions of the system are coupled to different degrees of freedom of the environment does not produce the usual quantum decoherence effects.

References

1 J. Fell, G. Fernandez, P. Klaver, C. E. Elger, P. Fries, Is Synchronized Neural Gamma Activity Relevant for Selective Attention? *Brain Res. Rev.* **42**, 265–72 (2003). A. K. Engel, P. Fries, and W. Singer, Dynamic Predictions: Oscillations and Synchrony in Top-processing, *Nat. Rev. Neurosci.* **2**, 704–716 (October 2001) doi:1038/35094565.

2 H. P. Stapp, *Mind, Matter, and Quantum Mechanics* (Springer, Berlin, Heidelberg, New York, 1993/2004). H. P. Stapp, *Mindful Universe: Quantum Mechanics and the Participating Observer* (Springer, Berlin, Heidelberg, New York, 2007). H. P. Stapp, Quantum Interactive Dualism: An Alternative to Materialism, *J. Consc. Studies*. **12**, no. 11, 43–58 (2005). [http://www-physics.lbl.gov/~stapp/stappfiles.html] J. M. Schwartz, H. P. Stapp, and M. Beauregard, Quantum Theory in Neuroscience and Psychology: A Neurophysical Model of the Mind/Brain Interaction. *Phil. Trans. Royal Soc.* **B 360** (1458) 1306 (2005). H. P. Stapp, Physicalism versus Quantum Mechanics (2008), http://www-physics.lbl.gov/~stapp/stappfiles.html.

3 H. D. Zeh, On the Interpretation of Measurement in Quantum Theory, *Found. Phys.* **1**, 69–76 (1970). E. Joos and H. D. Zeh, The Emergence of Classical Properties through Interaction with the Environment, *Z. Phys.* **B5**, 223–43 (1985). D. Giulini, E. Joos, C. Kieffer, J. Kupsch, I.-O. Stamatescu, and H. D. Zeh, *Decoherence and the Appearance of a Classical World in Quantum Theory* (Springer, Berlin, Heidelberg, New York, 1996). A. J. Leggett, Macroscopic Quantum Systems and the Quantum Theory of Measurement, *Supp. Prog. Theor. Phys.* **69**, 80–100 (1980). M. Tegmark, Importance of Quantum Decoherence in Brain Process, *Phys. Rev.* **E 61**, 4194–206 (2000).

4 Ref. 1.

5 J. Von Neumann, *Mathematical Foundations of Quantum Mechanics* (Princeton University Press, Princeton, 1955), chap. VI, p. 417.

6 W. Heisenberg, *Physics and Philosophy* (Harper and Row, New York, 1958), chap. III, The Copenhagen Interpretation.

7 Ref. 5.

8 Ref. 2.
9 J. R. Klauder and B. Skagerstam, *Coherent States* (World Scientific, Singapore, 1985), pp. 12, 20. V. Bargmann. P. Butera, L. Giradello, and J. R. Klauder, On the Completeness of the Coherent States, *Rep. Math. Phys.* **2**, 221–8 (1971). A. M. Perelomov, On the Completeness of a System of Coherent States, *Teor. Mat. Fiz.* **6**, 213–24 (1971), pp. 156-64 [English translation].

Part V

Appendices

A Mathematical Model

Let the thoughts be numbered and let thought number n and its associated pattern of neural excitation be labeled by $T(n)$. It has the structure

$$T(n) = T(n, 1) * T(n, 2) * \cdots * T(n, N),$$

where

$$T(n, i) = [T(n, i, 1) + T(n, i, 2) + \cdots + T(n, i, N_{ni})].$$

The top line represents James's marching column of serial components, with $T(n, 1)$ the fresh arrival and $T(n, N)$ the one just fading out. The $T(n, i, j)$s in the square bracket are the (parallel) components of $T(n, i)$. Each $T(n, i, j)$ tends to evolve in one serial step into the collection of protothoughts $T'(m)$ that it symbolizes:

$$T(n, i, j) \rightarrow \sum_m e(n, i, j; m) T'(m).$$

The incidence matrix $e(n, i, j; m)$ is an array of zeros and ones. A protothought is a brain pattern having the same kind of structure as a brain pattern $T(n)$.

Evolution under the mechanical, unconscious brain process generates

$$T(n) \rightarrow U_n(T(n)),$$

where

$$U_n(T(n)) = P(n + 1, 1) S(n + 1, 1) + P(n + 1, 2) S(n + 1, 2) + \cdots.$$

Each $S(n + 1, m)$ is a different possible $T(n + 1)$. The Heisenberg event gives

$$T(n + 1) = S(n + 1, m)$$

with probability $P(n + 1, m)$. This is the general format. But the operator U_n must be spelled out in detail.

The action of U_n is

$$U_n\left(T(n)\right) = D_n\left(\sum_{i=1}^{N}\sum_{j,m} e(n, i, j; m)\, T'(m)\right).$$

The argument of D_n is the collection of $T'(m)$ s into which $T(n)$, by it-self, would tend to evolve in one serial step. But the brain process has other sources, for example the senses. The information from these sources becomes, after due process, cast into a sum of $T'(m)$ s that I call

$$E_n = \sum_m e(n, m)\, T'(m).$$

The action of D_n is this: it first adds to its argument the collection E_n just described. This enlarged collection C_n is the set of $S(n+1, m)$ s. Thus only the probabilities $P(n+1, m)$ remain to be specified.

It may happen that certain of the $S(n+1, m)$ s occur more than once in C_n. Because the brain process is occurring in real time, through the actions of neural elements that respond in a highly nonlinear way to the strength of the stimulus, any pattern of excitations that is being stimulated to occur in several ways should have a strongly enhanced probability of occurring sooner. We assume that those patterns $T'(m)$ s that would have been activated later are strongly suppressed. Then the weighting $P(n+1, m)$ should be a rapidly increasing function of the number of ways that $S(n+1, m)$ can be formed out of the $T'(m)$ s in C_n.

The brain itself is a complex organ, and the simple model described above is not meant to be taken literally. But it gives the main idea of how consciousness is supposed to fit into brain dynamics. The serial structure allows the thought to grasp an *evolving* situation, and the rapid increase in the probability factor $P(n+1, m)$ mentioned above provides for efficient associative memory, by virtue of the fact that the potential thought with the greatest number of paths from the current thought will have a far greater probability of occurring.

Two examples will explain what I mean.

I wake one day and immediately see a book that is thick and blue. My first thought of the day is my thought number n, and it has fresh arrival

$$T(n, 1) = [\text{book} + \text{thick} + \text{blue}],$$

and no other serial component. Each of the three parallel components tends to generate the class of protothoughts it symbolizes. Thus the collection C_n, written as a sum, is

$$C_n = \sum_{j=1}^{3} \sum_{m} e(n, 1, j; m) \, T'(m),$$

where the three values of j correspond to "book", "thick", and "blue". C_n will be the collection of all protothoughts of books, thick things, and blue things. My protothought of my volume of Plato appears three times in this collection, but all other protothoughts appear only once or twice. Because of the quickness of its appearance the protothought of the book *Plato* will get almost all the weight: $P(n + 1, Plato) \approx 1$. Hence, with near certainty,

$$T(n + 1) = Plato.$$

This analog retrieval is a one-step process.

The $T'(m)$s must be *allowed representations* of the self and surroundings. This structure of allowed representations must itself have become etched into the structure of the brain. But suppose that I have by now a properly conditioned brain, and, for simplicity, that each "allowed world" is pictured, psychologically, as a point moving with constant velocity on a straight line in two-dimensional space. Consider the following situation. My initial sighting at psychological time $t = n$ is of the world at the origin:

$$T(n, 1) = [(x, y, t)] = [(0, 0, n)].$$

This datum tends to generate the collection of $T'(m)$s that represent straight spacetime lines that contain the point $(0, 0, n)$. There is no further clue until time $t = n + 6$. Then there is a fresh datum

$$T(n + 6, 1, 1) = [(2, 2, n + 6)].$$

The original datum will have moved to $T(n + 6, 7, 1)$. Only one possible world-image is contained both in the collection of straight spacetime lines that pass through $(2, 2, n+6)$ and also in the collection of straight spacetime lines that pass though $(0, 0, n)$. The brain's representation of this cross-referenced $T'(m)$ will occur sooner than any other in the set of $S(n+7, k)$s, and this $T(m)$ will therefore with high probability become $T(n+7)$. Notice that the entire picture of the moving world, during the interval covered by the thought $T(n + 1)$, snaps into place all at once at time $n + 7$.

These two examples illustrate how two fundamental features of efficient mind/brain functioning come out automatically. These are the rapid access of associative memory and the snapping into place at some psychological instant of an image of an entire *duration* of the self and surroundings *in flux*. The brain process constructs an entire scenario, filling in, in general, both a reconstructed past and an extrapolated future.

Glossary

actual A status of being, distinguished from possible: what is *possible* might come into being, whereas what is *actual* has definitely come into being.

appearances The forms that appear in our conscious thoughts when we direct our attention to particular aspects of the world around us. Also, the occurrences in our thoughts of those forms.

atomic phenomena Appearances that seem to reveal the occurrence of atomic events. An example is the appearance of the cloud-chamber tracks that we attribute to particles emerging from an atomic collision.

Bohr Niels Bohr, the Danish physicist who was the main guiding figure in the creation of quantum theory, and who labored to formulate a coherent philosophy that would allow scientists to pursue the development and application of that theory without becoming enmeshed in metaphysical questions that seemed to arise in connection with the theory, but whose answers were not needed for many applications of the theory.

classical concepts The ideas that characterize classical physics. The first main idea of classical physics is that the physical world can be represented in physical theory by a *simple aggregation* of simple properties, each of which can be assigned to a point (or small region) in spacetime. This is the local-reductionistic hypothesis. The second main idea is that, for example, a chair occupies a certain location in a room rather than being smeared out all over the room. A probability distribution for a chair could be smeared out all over the room: the chair might just as likely be in one place as another. But a chair itself, according to classical ideas, is definitely in one single place. On the other hand, an atom, according to orthodox quantum-theoretical ideas, is completely represented by a "probability distribution", and hence it can, in some sense, be in two places at once. However, chairs are seemingly built of atoms. This leads to the basic interpretational problem in quantum theory, which is to reconcile the facts that (1) chairs—and also pointers on measuring devices — appear to be confined to definite locations, whereas (2) the collections of atoms from which they are built are not always similarly constrained by the theory.

complementarity A key feature of the philosophy of Niels Bohr. The idea is that a system can be probed in different complementary ways, and that the properties that emerge under the action of different probings, while all equally essential to a full characterization of the attainable knowledge pertaining to the system, may not be representable as properties simultaneously possessed by the system: the probing action can be an essential part of making a property definite.

consciousness That luminescent presence of coming-into-beingness that constitutes our inner world of experience. It is present during our wakeful states, and during our dreams, but is extinguished during dreamless sleep, and in the state of unconsciousness induced by a severe blow to the head. It is not characterized by our behavior, because it can be present when the body is motionless, yet absent in the state of somnambulism.

Copenhagen interpretation A viewpoint developed by Bohr, Heisenberg, Pauli and others about how quantum theory is to be used and understood. The first main point, in the words of Bohr, is that "the task of science is both to extend the range of our experience and reduce it to order . . .". Note that the focus is on "our experience", rather than on nature herself: the task of science is to understand the structure of our experience, not the structure of some unexperienced "external reality". The second main point, again in the words of Bohr, is that ". . . the appropriate physical interpretation of the symbolic quantum-mechanical formalism amounts only to predictions, of determinate or statistical character, pertaining to individual phenomena appearing under conditions defined by classical physical concepts." That is, the mathematical procedure of quantum theory is merely a *tool* that we use to form expectations about what will appear to observers under certain special kinds of conditions. These conditions are to be described in terms of the concepts of classical physics. Classical physics is the fundamentally inaccurate theory that quantum theory displaced. The latent inconsistency in using in the formulation of quantum theory a theory that is both incompatible with quantum theory itself, and fundamentally incorrect, is an awkwardness appreciated as much by the originators of the Copenhagen interpretation as by its detractors. The third key feature is "Complementarity", which is described elsewhere in this glossary.

Descartes René Descartes, called the father of modern philosophy, who broke free, to some extent, from the dogmas of the scholastic philosophy of the Middle Ages, and, by pursuing the method of systematic doubt, arrived at what appeared to him to be an unquestionably true proposition, "I think, therefore I am", which was to serve as a foundation for a new philosophy not based upon appeal to authority. Thinking, in his philosophy, came before matter, and was logically separate from it. This conceptual separation freed physics from the need to consider thoughtlike things, and allowed it to focus exclusively on mathematical properties tied to geometry, a path that led to the triumphantly successful classical mechanics of Isaac Newton.

deterministic theory   A theory in which properties can take values at various times, and in which the set of all the values taken by properties at some early set of times *fixes completely* all values to be taken by properties at later times.

Einstein   Albert Einstein, the most celebrated and important scientist since Isaac Newton, and creator of the special and general theories of relativity. He made very important contributions to the development of quantum theory, but, in opposition to the "orthodox" philosophy of Bohr, held that science should endeavor to construct theories about the world itself, not merely our experience of it. In spite of Einstein's enormous prestige, his views on this matter were for many years dismissed as outdated by most quantum physicists. But contemporary efforts to extend quantum theory to new domains are now turning increasing numbers of physicists to the view that Bohr's philosophy is too limiting on theoretical creativity, just as Einstein had maintained.

EPR   The initials of (Albert) Einstein, (Boris) Podolsky, and (Nathan) Rosen, the authors of a famous 1935 scientific paper entitled "Can Quantum-Mechanical Description of Physical Reality Be Considered Complete?" Within the mathematical structure of quantum theory the position and the velocity of an electron cannot both be well defined simultaneously: if one of these quantities has a definite value the other cannot. The EPR paper noted that within quantum theory one can set up an experimental situation that in principle allows one to predict with certainty either the position or the velocity of a first electron by measuring instead either the position or the velocity of a second electron. EPR argued that measurements performed on the second electron cannot disturb the properties of the first electron, and hence that the first electron must have, simultaneuosly, a well-defined position and a well-defined velocity. But quantum theory cannot describe such a situation.

event   An occurrence that covers a limited time span: i.e., that ends "shortly" after it begins. A *quantum event* is the sudden change of the quantum state from one form to another. This event is also called the quantum jump, the collapse of the wave function, or the reduction of the wave packet.

experience   The collection of events, or happenings, that constitute our conscious mental life.

functionalism   An attempt to circumvent the conflict between the unity of consciousness and the precepts of classical physics by assigning ontological status to physical structures on the basis of what they can *do*, in an appropriate context, rather than on the basis of what they *are*, intrinsically.

Heisenberg   Werner Heisenberg, the chief inventor of quantum theory. The main innovation of that theory is that physical properties, such as the position of a particle, or its momentum, are not represented by mere passive numbers. They are represented instead by "actions": they are represented by "operators" that *act* on states, and change these states to other states. Each *act of measurement* is also represented by an "action". This action changes the state of the system acted upon into a new state, which has the following feature: the action upon this new state of the operator that represents the property being measured is simply to multiply this state by a number. This number is the *value* given to that property by the act of measurement: it is the *result*, or the *outcome*, of the measurement.

Heisenberg ontology   An ontology suggested by the later writings of Heisenberg. I may have gone beyond the explicit statements of Heisenberg by specifying that his actual events occur not only in true measurement situations, in which there is a human observer of some external device, but equally in all physically similar situations, regardless of whether a human observer is present or not. This is in line with his assertion that the actual events occur at the level of the devices, not at the level of the registration of the result in the mind of the observer. I have also tightened the connection between Heisenberg's ontological statements and the mathematical structure of quantum theory by giving to the latter an objective ontological status that goes beyond that ascribed to it by the Copenhagen interpretation.

James   William James, professor of psychology at Harvard, and author of, among other works, the voluminous *Principles of Psychology*, which contains an enormous wealth of data and logical analysis pertaining to the nature of consciousness. In later life James was (deservedly, according to Bertrand Russell) the recognized leader of American philosophy. He separated the empirical facts about consciousness from the notion of a "knower", and his thinking presages quantum theory in at least three ways: (1) He was one of three main protagonists for a practical or "pragmatic" theory of truth according to which the truth of an idea lies ultimately in how well it works in our lives. This idea is closely connected to Bohr's Copenhagen interpretation of quantum theory. (2) He recognized clearly the incompatibility of the unity of consciousness with the reductionistic character of classical physics, and hence the need for a new idea about the nature of matter. (3) He eventually came to hold the view that there is only one kind of "primal stuff" out of which everything in the universe is made, and this stuff he called "pure experience". This view is not far out of line with the idea that the primal stuff of the universe is what is represented in quantum theory as the evolving state of the universe, and that certain changes of this state are the counterparts, within this theory, of conscious human experiences.

locality   The property of classical physical theories that forbids any instantaneous nearby effect of a faraway cause.

many-worlds interpretation   An interpretation of quantum theory that holds that the basic equation of motion in the theory, the Schrödinger equation, always holds, and that the basic quantity of the theory, the "probability distribution", is not merely a computational tool, nor merely a quantity that determines the probability for some *event* to occur. Rather, the quantum probability distribution is claimed to be also a representation of the total physical situation itself. Just as the probability distribution in classical physics evolves naturally, in the course of time, into a collection of parts representing different possible observable courses of events, each with a statistical weight, so likewise does the quantum probability distribution normally split into a set of branches representing the different possible courses of events that *might* appear to a human observer. For example, the probability distribution representing a world containing "Schrödinger's cat" evolves into two branches, one representing a world containing an alive cat and one representing a world containing a dead cat. According to the many-worlds interpretation both courses of events actually occur in nature: eventually there will be, within the fullness of nature, both an alive version of the cat and a dead version of this cat, existing in the same box. By virtue of the Schrödinger equation the two versions of the cat that exist together in the box do not affect each other: each evolves as if the other were not there. If some member of a community of communicating observers looks into the box, then this whole community will split into two independent communities, one containing the alive cat and the other containing the dead cat.

This interpretation appears, superficially, to be confounding the many *possible* courses of events in a statistical ensemble of possibilities with the one course of events that actually occurs.

materialism   An opinion about the nature of all things that is distinguished from idealism and dualism. Materialism is the view that everything is made of matter, and, in particular, that the ideas, thoughts, and feelings that make up our conscious inner lives are merely complex combinations of the same sort of ingredients, namely atoms and local fields, that rocks and mountains are made of. Idealism is the converse opinion that the fundamental components from which everything is made are elements like our conscious thoughts, and that therefore rocks are in some sense built out of ideas. Dualism is the intermediate opinion that rocklike things and idealike things are two fundamentally different kinds of basic elements, and that the totality of nature is built out of a combination of these two distinct kinds of things. This entire controversy about what the world is made of seems to dissolve into an ill-informed verbal dispute if the universe conforms to the quantum description of it. For in this description there is only one "stuff", namely the evolving quantum state of the universe, but this stuff has two very different modes of dynamical evolution: the smooth and the abrupt. The smooth development is matterlike in the sense that it is locally lawful, whereas the abrupt change is idealike in the sense that it injects free choice. On the other hand, the smooth development is idealike in that it represents the evolution of merely the potentialities and probabilities for the actual things. And these actual things are not enduring matterlike objects, but rather certain fleeting happenings, the quantum jumps. Thus the physical world, as described by quantum theory, is an intricately interwoven combination of qualities of the kind that we usually associate with the concepts of mind and matter. The thrust of the present work is that when this quantum description of matter is applied to the human brain it accounts naturally also for the prime idealike quality, human consciousness. This outcome can be viewed as a vindication of materialism, since the theory of matter automatically subsumes consciousness. Alternatively, it can be viewed as confirmation of an idealism of the kind that James called "radical empiricism", because the single primal stuff, which is represented in the theory by the evolving quantum state of the universe, would probably better be called "pure experience" than "pure matter", owing to its nonsubstantive nature, and the fact that a certain aspect of the behavior of this primal stuff is human conscious experience. Yet in spite of its basically idealike nature the primal quantum stuff has certain mathematical qualities that resemble those of matter. Hence this stuff might be called "mind/matter". Better, however, is "mind/math", for this replacement of "matter" by "math" emphasizes that what is present in nature is not the substantiveness, or rocklike quality, that we often associate with the word "matter" but rather a partial conformity to mathematical rules. If mathematics is deemed to belong to the category of "mind", then the primal quantum stuff could be called simply *mind*, but only in a sense that allows mind to include not only human conscious experience but also the mathematical aspects of the way this primal stuff evolves. Showing how these two apparently disparate parts of nature can hang together in a rationally coherent way is the aim of the present work.

matter   In this work *matter* denotes those aspects of nature that can be represented as a collection of properties that (1) are localizable at or near spacetime points, (2) evolve continuously according to deterministic equations of motion, and (3) carry energy and momentum.

measurement   The operation of eliciting an observable phenomenon that is an indicator of some aspect or property of a system.

mind   In this work *mind* denotes a category that includes conscious human thoughts, and things that appear to be like them.

ontological   Having to do with what exists, and distinguished from epistemological, which means having to do with what we can know.

Pauli   Wolfgang Pauli, Nobel prize winning physicist, who was one of the principal founders of quantum theory, and was regarded as perhaps the most incisive thinker in that group.

phenomena   "Appearances", as distinguished from their causes: that which appears in conscious experience itself.

phenomenology   The study of experience.

probability   A conceptual tool whereby various possibilities are assigned "statistical weights", which, however, are given strict interpretations only in terms of idealized, unrealizable situations involving infinite numbers of instances. These "weights" carry a general intuitive meaning of "tendency to occur", or, if the statistical weight of the possibility is very close to unity, of "near certainty to occur".

process   The unfolding of nature: the coming-into-being-ness of the totality of which our conscious experience is a connected part.

quantum theory   A set of rules that allow scientists to make predictions about what will appear to human observers under certain specified kinds of conditions. These predictions have been validated in an enormous number of diverse situations. No unambiguous prediction of quantum theory has ever been shown to be false. Quantum theory is distinguished from classical physics, which works well for systems that do not depend critically upon what is going on at the atomic level.

representation   A structure that, within a certain context, stands for, or takes the place of, another structure. The first structure is said to *represent* the second. The context within which such a representation can occur must include some mechanism of interpretation, which can recognize the representation and respond to it in a characteristic way.

Schrödinger   Erwin Schrödinger, the inventor of a form of the basic equation of motion in quantum theory, and of the particular mathematical representation of the quantum state that his equation governs. The Schrödinger equation is a transcription, into a new form, of a corresponding law of motion of classical physics, and, like the latter, it generates a continuous deterministic evolution of the state of the system. In another way of expressing the theory the Schrödinger equation becomes the Heisenberg equations.

Schrödinger's cat   The cat in an imaginary experiment discussed by Erwin Schrödinger. This cat is placed in a black box with a mechanism that will release a pellet of cyanide gas if the decay of a radioactive nucleus is detected by a certain detecting device. According to the Schrödinger equation, the whole system, as it is represented in quantum theory, will evolve into a system that is a superposition of two parts: in one part the cat is alive; in the other part the cat is dead. The behavior of the quantum representation of this system must be reconciled with the fact that only one of these two cats will appear in the experience of any actual human observer of the system.

statistical theory   A theory based upon the notion of chance: a theory in which each possibility is assigned a statistical weight, which is interpreted as the probability for this possibility to be, or to become, actual.

superposition   In a classical statistical analysis one contemplates a collection of possibilities. Each of these possibilities evolves in the course of time independently of the others, because the possibilities are combined only in our thoughts, not in reality. In quantum theory the various possibilities in the statistical analysis combine in a fundamentally different way. The word used to denote this different way of combining possibilities is "superposition". Different superposed possibilities both do and do not evolve independently of each other: if one considers the so-called probability amplitude, which is roughly the square root of the probability, then the various superposed parts do evolve independently of each other; but, owing to the need to square, the probabilities themselves do not enjoy this property. This peculiar behavior lies at the root of the difficulties that scientists and nonscientists alike have in coming to the belief that they "really understand" quantum theory.

time   Opening to change: the dimension of nature that allows for evolution. In quantum theory there are two different modes of evolution, and there are, correspondingly, two different kinds of time. The first is *process time*, or *actual time*. It is the time that is marked by the sequence of actual events— by the discrete succession of quantum jumps in the Heisenberg state of the universe. The second time is *Einstein time*, or *virtual time*. It is the time that is joined with space to form spacetime, and is the time associated with the mathematical equations of motion, namely the Heisenberg equations for the evolution of the operators that correspond to the physical quantities of classical physics. This latter evolution is *virtual* in the sense that it is the development not of the actual things themselves, but of only the potentialities and probabilities for the actual things.

unity of consciousness   A purported property of consciousness according to which each conscious thought is a complex entity that cannot be decomposed into a simple aggregation of simple components without destroying its essence. This unity, forcefully claimed by James, makes it impossible, in principle, to represent a conscious thought faithfully within classical physics, for the latter can represent faithfully only things that are in essence simple aggregates of simple local properties.

von Neumann quantum theory   This is a formulation in which the entire physical universe, including the bodies and brains of the conscious human participant/observers, is represented by the basic quantum state. The dynamics involves three processes. Process **I** is the choice on the part of the experimenter about how he will act. This choice is sometimes called the "Heisenberg choice", because Heisenberg emphasized strongly its crucial role in quantum dynamics. At the pragmatic level it is a "free choice", because it is controlled, at least at the practical level, by the conscious intentions of the experimenter/participant: neither the Copenhagen nor von Neumann formulations specify the causal origins of this choice, apart from the conscious intentions of the human agent. Process **II** is the quantum analog of the equations of motion of classical physics, and like its classical counterpart is local (i.e., via contact between neighbors) and deterministic. This process is constructed from the classical one by a certain quantization procedure, and is reduced back to the classical process by taking the classical approximation. It normally has the effect of expanding the microscopic uncertainties demanded by the Heisenberg uncertainty principle into the macroscopic domain: the centers of large objects are smeared out over large regions of space. This conflict with conscious experience is resolved by invoking Processes **I** and **III**. Process **III** is sometimes call the "Dirac choice". Dirac called it a "choice on the part of Nature". It can be regarded as Nature's answer to a question effectively posed by the Process **I** choice made by the experimenter. This posed question is: Will the intended consequences of the action that the agent chooses to perform actually be experienced? (e.g., Will the Geiger counter be observed to be placed in the intended place? And, if so, Will the specified action (e.g., firing) of that device be observed to occur? Processes **I** and **III** act on the variables that specify the body/brain of the agent. The "Yes" answer actualizes the neural correlates of the intended action or associated feedback.

von Neumann/Stapp theory   The von Neumann theory, as described above, together with the assumption that the causal efficacy of conscious will arises from the activation, by willful effort, of a rapid sequence of Process **I** actions, which triggers a quantum Zeno effect, which holds the brain state in an associated subspace longer than what Process **II** would otherwise allow.

Further References

B. J. Baars, *A Cognitive Theory of Consciousness* (Cambridge University Press, Cambridge, 1988).

D. Bohm, *Phys. Rev.* **85**, 166–193 (1952).

W. Calvin, *The Cerebral Symphony* (Bantam Books, New York, 1990).

W. Calvin, *The Ascent of Mind* (Bantam Books, New York, 1991).

P. S. Churchland, *Neurophilosophy: Toward a Unified Science of the Mind/Brain* (MIT Press, Cambridge MA, 1986).

D. C. Dennett, *Consciousness Explained* (Little, Brown, & Co., New York, 1991).

J. C. Eccles, *Proc. Roy. Soc. London, Series B* **227**, 411–428 (1986).

G. M. Edelman, *Bright Air, Brilliant Fire: On the Matter of Mind* (Basic Books, New York, 1992).

L. Hardy, *Phys. Rev. Lett.* **68**, 2981–2988 (1992).

W. Heisenberg, *Physics and Philosophy* (Harper and Row, New York, 1958).

M. Lockwood, *Mind, Brain and the Quantum: The Compound "I"* (Basil Blackwell, Cambridge MA, 1989).

R. Penrose, *The Emperor's New Mind* (Oxford University Press, New York, 1989).

K. R. Popper and J. C. Eccles, *The Self and Its Brain* (Springer-Verlag, Berlin Heidelberg, 1977).

G. Ryle, *The Concept of Mind* (Barnes and Noble, New York, 1949).

J. R. Searle, *Minds, Brains, and Science* (Harvard University Press, Cambridge MA, 1984).

J. R. Searle, *The Rediscovery of the Mind* (MIT Press, Cambridge MA, 1992).

J. B. Watson, *Psychol. Rev.* **20**, 158–195 (1963).

Index

THE FRONTIERS COLLECTION

Series Editors:
A.C. Elitzur M. Schlosshauer M.P. Silverman J. Tuszynski R. Vaas H.D. Zeh